Six-Minute Solutions

for Civil PE Exam

Water Resources and Environmental Problems

R. Wane Schneiter, PhD, PE

Professional Publications, Inc. • Belmont, California

Benefit by Registering This Book with PPI

- Get book updates and corrections.
- Hear the latest exam news.
- Obtain exclusive exam tips and strategies.
- Receive special discounts.

Register your book at **www.ppi2pass.com/register**.

Report Errors and View Corrections for This Book

PPI is grateful to every reader who notifies us of a possible error. Your feedback allows us to improve the quality and accuracy of our products. You can report errata and view corrections at **www.ppi2pass.com/errata**.

SIX-MINUTE SOLUTIONS FOR CIVIL PE EXAM
WATER RESOURCES AND ENVIRONMENTAL PROBLEMS

Current printing of this edition: 4

Printing History

edition number	printing number	update
1	2	Minor corrections.
1	3	Minor corrections.
1	4	Minor corrections. Copyright update.

Copyright © 2012 by Professional Publications, Inc. (PPI). All rights reserved. No part of this publication may be reproduced, stored in a retrieval system, or transmitted, in any form or by any means, electronic, mechanical, photocopying, recording, or otherwise, without the prior written permission of the publisher.

Printed in the United States of America.

PPI
1250 Fifth Avenue, Belmont, CA 94002
(650) 593-9119
www.ppi2pass.com

ISBN: 978-1-59126-139-1

Library of Congress Control Number: 2008927585

Table of Contents

ABOUT THE AUTHOR v

PREFACE AND ACKNOWLEDGMENTS vii

INTRODUCTION
    Exam Format ix
    This Book's Organization x
    How to Use This Book x

REFERENCES ix

BREADTH PROBLEMS
    Hydraulics—Closed Conduit 1
    Hydraulics—Open Channel 2
    Hydrology 4
    Wastewater Treatment 6
    Water Treatment 7

DEPTH PROBLEMS
    Hydraulics—Closed Conduit 9
    Hydraulics—Open Channel 10
    Hydrology 10
    Groundwater and Well Fields 13
    Wastewater Treatment 15
    Water Quality 18
    Water Treatment 22
    Engineering Economics 24

BREADTH SOLUTIONS
    Hydraulics—Closed Conduit 25
    Hydraulics—Open Channel 29
    Hydrology 33
    Wastewater Treatment 37
    Water Treatment 39

DEPTH SOLUTIONS

Hydraulics—Closed Conduit 43

Hydraulics—Open Channel 46

Hydrology 50

Groundwater and Well Fields 56

Wastewater Treatment 60

Water Quality 68

Water Treatment 73

Engineering Economics 80

About the Author

R. Wane Schneiter, PhD, PE, is the Deputy Superintendant for Academics and Dean of the Faculty at the Virginia Military Institute. He joined the VMI faculty in 1990. Prior to his appointment as dean he served as the head of the Civil and Environmental Engineering Department and held the Benjamin H. Powell Jr. '36 Professorship in Engineering. His professional engineering activities have included consulting services on environmental issues to a wide array of public and private clients. He holds BS and PhD degrees in Civil Engineering from Utah State University.

Preface and Acknowledgments

The Principles and Practice of Engineering examination (PE exam) for civil engineering, prepared by the National Council of Examiners for Engineering and Surveying (NCEES), is developed from sample problems submitted by educators and professional engineers representing consulting, government, and industry. PE exams are designed to test examinees' understanding of both conceptual and practical engineering concepts. Problems from past exams are not available from NCEES nor any other source. However, NCEES does identify the general subject areas covered on the exam.

The topics covered in *Six-Minute Solutions for Civil PE Exam Water Resources and Environmental Problems* coincide with those subject areas identified by NCEES for water resources and environmental engineering depth module of the civil PE exam. These problem topics are hydraulics, hydrology, groundwater and well fields, water and wastewater treatment, water quality, and engineering economics.

The problems presented in this book are representative of the type and difficulty of problems you will encounter on the PE exam. They were previously published in either *Six-Minute Solutions for Civil PE Exam Problems: Water Resources* or *Six-Minute Solutions for Civil PE Exam Environmental Problems*, and have been reorganized and modified to follow the current exam specifications. The book's problems are both conceptual and practical, and they are written to provide varying levels of difficulty. Though you probably won't encounter problems on the exam exactly like those presented here, reviewing these problems and solutions will increase your familiarity with the exam problems' form, content, and solution methods. This preparation will help you considerably during the exam.

Problems and solutions have been carefully prepared and reviewed to ensure that they are appropriate and understandable, and that they were solved correctly. If you find errors or discover a more efficient way to solve a problem, please bring it to PPI's attention so your suggestions can be incorporated into future editions. You can report errors and keep up with the changes made to this book, as well changes to the exam, by logging on to PPI's website at **www.ppi2pass.com/errata**.

Thank you to the many persons in the editorial and production departments at PPI who contributed to the successful publication of the second edition of this book. They are an enjoyable group to work with, are thorough and professional, and are dedicated to providing the best possible publication.

R. Wane Schneiter, PhD, PE

Introduction

EXAM FORMAT

The Principles and Practice of Engineering examination (PE exam) in civil engineering is an eight hour exam divided into a morning and an afternoon session. The morning session is known as the *breadth* exam, and the afternoon is known as the *depth* exam.

The morning session consists of 40 problems from all five civil engineering subdisciplines. As the "breadth" designation implies, morning session problems are general in nature and wide-ranging in scope. Topics and the approximate distribution of problems on the morning session of the exam are as follows.

Construction: approximately 20% of exam problems

Earthwork construction and layout; estimating quantities and costs; scheduling; material quality control and production; temporary structures

Geotechnical: approximately 20% of exam problems

Subsurface exploration and sampling; engineering properties of soils and materials; soil mechanics analysis; earth structures; shallow foundations; earth retaining structures

Structural: approximately 20% of exam problems

Loadings; analysis; mechanics of materials; materials; member design

Transportation: approximately 20% of exam problems

Geometric design

Water Resources and Environmental: approximately 20% of exam problems

Hydraulics—closed conduit and open channel; hydrology; wastewater treatment; water treatment

The afternoon session allows the examinee to select a depth exam module from one of the five subdisciplines. The 40 problems in the afternoon session require more specialized knowledge than those in the morning session.

All problems from the morning and afternoon sessions are multiple choice. They include a problem statement with all required defining information, followed by four logical choices. Only one of the four options is correct. Nearly every problem is independent of the others, so an incorrect choice on one problem typically will not carry over to subsequent problems.

Topics and the approximate distribution of problems on the afternoon session of the civil water resources and environmental exam are as follows.

Hydraulics—Closed Conduit: approximately 15% of exam problems

Energy and/or continuity equation; pressure conduit; closed pipe flow equations; friction and/or minor losses; pipe network analysis; pump application and analysis; cavitation; transient analysis; flow measurement—closed conduits; momentum equation

Hydraulics—Open Channel: approximately 15% of exam problems

Open-channel flow; culvert design; spillway capacity; energy dissipation; stormwater collection including stormwater inlets, gutter flow, street flow, storm sewer pipes; flood plain/floodway; subcritical and supercritical flow; flow measurements—open channel; gradually varied flow

Hydrology: approximately 15% of exam problems

Storm characterization; storm frequency; hydrographs application; hydrograph development and synthetic hydrographs; rainfall intensity, duration, and frequency (IDF) curves; time of concentration; runoff analysis; gauging stations including runoff frequency analysis and flow calculations; depletions; sedimentation; erosion; detention/retention ponds

Groundwater and Well Fields: approximately 7.5% of exam problems

Aquifers; groundwater flow; well analysis; groundwater control; water quality analysis; groundwater contamination

Wastewater Treatment: approximately 15% of exam problems

Wastewater flow rates; unit operations and processes; primary treatment; secondary clarification; chemical treatment; collection systems; National Pollutant Discharge Elimination System (NPDES) permitting; effluent limits; biological treatment; physical treatment; solids handling; digesters; disinfection; nitrification and/or denitrification; operations; advanced treatment; beneficial reuse

Water Quality: approximately 15% of exam problems

Stream degradation; oxygen dynamics; risk assessment and management; toxicity; biological contaminants; chemical contaminants; bioaccumulation; eutrophication; indicator organisms and testing; sampling and monitoring

Water Treatment: approximately 15% of exam problems

Demands; hydraulic loading; storage; sedimentation; taste and odor control; rapid mixing; coagulation and flocculation; filtration; disinfection; softening; advanced treatment; distribution systems

Engineering Economics: approximately 2.5% of exam problems

Life-cycle modeling; value engineering and costing

For further information and tips on how to prepare for the civil water resources and environmental PE exam, consult the *Civil Engineering Reference Manual* or PPI's website, www.ppi2pass.com.

THIS BOOK'S ORGANIZATION

Six-Minute Solutions for Civil PE Exam Water Resources and Environmental Problems is organized into two sections. The first section, Breadth Problems, presents 31 water resources engineering problems of the type that would be expected in the morning part of the civil engineering PE exam. The second section, Depth Problems, presents 69 problems representative of the afternoon part of this exam. The two sections of the book are further subdivided into the topic areas covered by the water resources exam.

Most of the problems are quantitative, requiring calculations to arrive at a correct solution. A few are non-quantitative. Some problems will require a little more than six minutes to answer and others a little less. On average, you should expect to complete 80 problems in 480 minutes (eight hours), or spend six minutes per problem.

HOW TO USE THIS BOOK

In *Six-Minute Solutions for Civil PE Exam Water Resources and Environmental Problems*, each problem statement, with its supporting information and answer choices, is presented in the same format as the problems encountered on the PE exam. The solutions are presented in a step-by-step sequence to help you follow the logical development of the correct solution and to provide examples of how you may want to approach your solutions as you take the PE exam.

Each problem includes a hint to provide direction in solving the problem. In addition to the correct solution, you will find an explanation of the faulty solutions leading to the three incorrect answer choices. The incorrect solutions are intended to represent common mistakes made when solving each type of problem. These may be simple mathematical errors, such as failing to square a term in an equation, or more serious errors, such as using the wrong equation.

To optimize your study time and obtain the maximum benefit from the practice problems, consider the following suggestions.

1. Complete an overall review of the problems and identify the subjects that you are least familiar with. Work a few of these problems to assess your general understanding of the subjects and to identify your strengths and weaknesses.

2. Locate and organize relevant resource materials. As you work problems, some of these resources will emerge as more useful to you than others. These are what you will want to have on hand when taking the PE exam.

3. Work the problems in one subject area at a time, starting with the subject areas that you have the most difficulty with.

4. When possible, work problems without utilizing the hint. Always attempt your own solution before looking at the solutions provided in the book. Use the solutions to check your work or to provide guidance in finding solutions to the more difficult problems. Use the incorrect solutions to help identify pitfalls and to develop strategies to avoid them.

5. Use each subject area's solutions as a guide to understanding general problem-solving approaches. Although problems identical to those presented in *Six-Minute Solutions for Civil PE Exam Water Resources and Environmental Problems* will not be encountered on the PE exam, the approach to solving problems will be the same.

Solutions presented for each problem may represent only one of several methods for obtaining a correct answer. Although we have tried to prepare problems with unique solutions, alternative problem-solving methods may produce a different, but nonetheless appropriate, answer.

References

The minimum recommended library for the civil exam consists of PPI's *Civil Engineering Reference Manual*. You may also find current editions of references similar to the following helpful in completing some of the problems in *Six-Minute Solutions for Civil PE Exam Water Resources and Environmental Problems*.

Aisenbrey, A. J., Jr., et al. *Design of Small Canal Structures*. Denver, CO: U.S. Department of Interior, Bureau of Reclamation, 1978.

Dean, J. A. *Lange's Handbook of Chemistry*. 16th ed. New York: McGraw-Hill, 2004.

Fetter, C. W. *Applied Hydrogeology*. 4th ed. New York: Macmillan, 2000.

———. *Contaminant Hydrogeology*. 2nd ed. Upper Saddle River, NJ: Prentice Hall, 1999.

Lide, D. R. *Handbook of Chemistry and Physics*. 83rd ed. Boca Raton, FL: CRC Press, 2002.

Linsley, R. K., et al. *Hydrology for Engineers*. 3rd ed. New York: McGraw-Hill, 1982.

———. *Water-Resources Engineering*. 4th ed. New York: McGraw-Hill, 1991.

Luthin, J. N. *Drainage Engineering*. Huntington, NY: RE Krieger, 1978.

Masters, G. M. *Introduction to Environmental Engineering and Science*. 3rd ed. Upper Saddle River, NJ: Prentice Hall, 2007.

McGhee, T. J. *Water Supply and Sewerage Engineering*. 6th ed. New York: McGraw-Hill, 1991.

Metcalf & Eddy, Inc. *Wastewater Engineering: Treatment, Disposal, and Reuse*. 4th ed. New York: McGraw-Hill, 2003.

Munson, B. R., et al. *Fundamentals of Fluid Mechanics*. 5th ed. New York: John Wiley & Sons, 2005.

Peavy, H. S., et al. *Environmental Engineering*. New York. McGraw-Hill, 1985.

Sawyer, C. N., P. L. McCarty, and G. F. Parkin. *Chemistry for Environmental Engineering*. 5th ed. New York: McGraw-Hill, 2002.

Sincero, A. P., and G. A. Sincero. *Environmental Engineering: A Design Approach*. Upper Saddle River, NJ: Prentice-Hall, 1996.

Viessman, W., Jr., et al. *Introduction to Hydrology*. 5th ed. Menlo Park, CA: Addison-Wesley, 2002.

Viessman, W., Jr., and M. J. Hammer. *Water Supply and Pollution Control*. 7th ed. Menlo Park, CA: Addison-Wesley, 2004.

Breadth Problems

HYDRAULICS—CLOSED CONDUIT

PROBLEM 1

Water flows at 4.2 ft/sec in a 2 in diameter pipe that is connected through a reducer to a 1.5 in diameter pipe. What is the flow velocity in the 1.5 in diameter pipe?

(A) 2.4 ft/sec
(B) 5.6 ft/sec
(C) 7.5 ft/sec
(D) 28 ft/sec

Hint: This is a continuity equation problem.

PROBLEM 2

Water flows by gravity between two tanks through 250 ft of 2 in diameter steel pipeline with screwed fittings. The pipeline includes 7 regular 90° elbows, 16 couples, 4 unions, 5 tees (flow-through line), and 2 gate valves. What is the total equivalent length of the pipeline?

(A) 110 ft
(B) 260 ft
(C) 350 ft
(D) 360 ft

Hint: How is equivalent length defined?

PROBLEM 3

Water flowing in a schedule-40 steel pipe divides at a tee as shown in the illustration. The characteristics of each pipe are summarized in the table. The flow rate in pipe A is 0.76 ft³/sec.

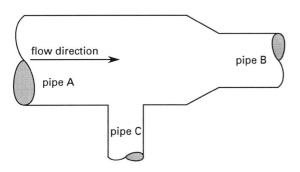

pipe	nominal diameter (in)	pressure (psig)
A	2.5	100
B	1.5	90
C	1	–

What is the flow rate in pipe C?

(A) 0.13 ft³/sec
(B) 0.22 ft³/sec
(C) 0.35 ft³/sec
(D) 0.76 ft³/sec

Hint: Use both the continuity and the energy equations.

PROBLEM 4

Water is pumped from an elevation of 3457 ft to an elevation of 3503 ft through 500 ft of 2 in schedule-80 steel pipe at a velocity of 5 ft/sec. The pipe includes 15 couples, eight 90° regular elbows, four 45° regular elbows, six tees (straight flow), and two globe valves. All fittings are standard pipe thread. The water temperature is 60°F. What is the total head loss in the pipe from all sources?

(A) 24 ft
(B) 43 ft
(C) 86 ft
(D) 270 ft

Hint: Head losses in this problem occur from two sources.

PROBLEM 5

Pressure is maintained at 100 psig in a fire hose with a nozzle diameter of 0.5 in. The nozzle coefficient is 0.98 and the water temperature is 70°F. What is the flow rate at the nozzle outlet?

(A) 13 gal/min
(B) 75 gal/min
(C) 240 gal/min
(D) 600 gal/min

Hint: This problem resembles a problem involving the discharge from an orifice at the bottom of a tank.

PROBLEM 6

A centrifugal pump operates at a speed of 1750 rpm and is rated at 850 gal/min for 78% efficiency and 180 ft of head. For constant efficiency and head, what is most nearly the flow rate if the pump is operated at 2200 rpm?

(A) 530 gal/min
(B) 680 gal/min
(C) 830 gal/min
(D) 1100 gal/min

Hint: Apply the principles of affinity.

PROBLEM 7

The net positive suction head (NPSH) characteristics of a pump are shown in the following illustration. The pump was selected to deliver 75 gal/min of water at 48°F. The friction losses between the water surface and the pump inlet are 6.7 ft.

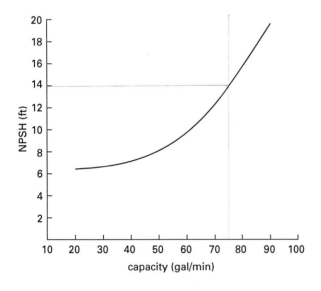

If the water surface elevation is 4573 ft above mean sea level (MSL), what is the maximum permissible elevation of the pump, if it is to be placed above the water surface?

(A) 4566 ft
(B) 4580 ft
(C) 4586 ft
(D) 4608 ft

Hint: Find an equation that relates net positive suction head to elevation.

PROBLEM 8

Water at 10°C is flowing 43 mm deep in a 100 mm diameter PVC pipe. The pipe is placed on a uniform 3% slope. What is most nearly the numeric value of the friction factor?

(A) 0.014
(B) 0.017
(C) 0.070
(D) 0.37

Hint: Find or develop an equation that relates the friction factor to known parameters.

HYDRAULICS—OPEN CHANNEL
PROBLEM 9

A radial gate is used to control flow into a wasteway turnout and prevent erosion of a section of downstream channel. The gate is operated partially open. The normal flow in the wasteway channel is 80 ft^3/sec with 4.3 ft of available head at the gate. The gate discharge coefficient is 0.72. What is the required area of the gate opening?

(A) 0.40 ft^2
(B) 0.83 ft^2
(C) 6.7 ft^2
(D) 54 ft^2

Hint: A partially open gate has the characteristics of a submerged orifice.

PROBLEM 10

What is the head loss per unit of length in a smooth earthen trapezoidal channel with a base width of 2 m, a water depth of 0.5 m, a water velocity of 3.2 m/s, and 1-to-1 side slopes?

(A) 0.00065 m/m
(B) 0.0040 m/m
(C) 0.013 m/m
(D) 0.024 m/m

Hint: Can any simplifying assumptions be made regarding head loss? What equations apply to head loss in open channel flow?

PROBLEM 11

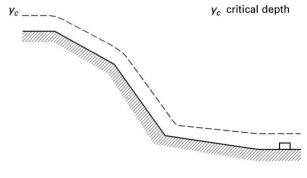

channel section

Which figure most likely represents the flow profile over the channel section shown?

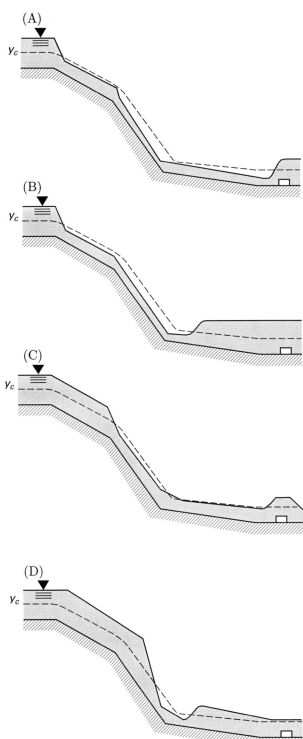

Hint: How is critical flow influenced by slope?

PROBLEM 12

A 1 m high, 27 m long box culvert placed at 2% slope sees an upstream water surface elevation of 1202.83 m and a downstream water surface elevation of 1202.38 m. The invert elevation of the culvert outlet is 1201.17 m. What is the culvert flow classification?

(A) type 3
(B) type 4
(C) type 5
(D) type 6

Hint: Sketch the culvert.

PROBLEM 13

A hydraulic jump with a stilling pool is selected to dissipate energy over a spillway prior to the water entering a natural river channel. The spillway is 2.5 m wide and the hydraulic jump occurs when the water depth at the toe of the spillway is 0.15 m for a flow of 2.93 m^3/s. What is most nearly the total head dissipated?

(A) 1.4 m
(B) 1.8 m
(C) 1.9 m
(D) 4.2 m

Hint: The solution involves the calculation of conjugate depths.

PROBLEM 14

A concrete-lined trapezoidal channel with an 8 m bottom width and 1-to-1 side slope will be used for storm water drainage. The maximum anticipated flow in the channel is 22 m^3/s. What uniform slope of the channel bottom is required to just produce critical flow at the maximum flow rate?

(A) 0.000060 m/m
(B) 0.00054 m/m
(C) 0.0019 m/m
(D) 0.25 m/m

Hint: Use equations for Froude number and channel cross-sectional area.

PROBLEM 15

A hydraulic jump occurs in a trapezoidal channel with a 4.2 m bottom width and 1-to-1 side slopes. The flow in the channel is 38 m^3/s and the water depth upstream of the jump is 0.74 m. What is the water depth downstream of the jump?

(A) 0.69 m
(B) 0.74 m
(C) 2.3 m
(D) 3.3 m

Hint: Use either conjugate depths or the specific force equation.

HYDROLOGY

PROBLEM 16

The cross section of a river channel is shown in the illustration. Velocity measurements for each section of river channel at the indicated depths are summarized in the table.

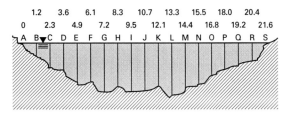

section	velocity at 0.2 depth (m/s)	velocity at 0.8 depth (m/s)	average depth (m)
AB	–	–	0.7
BC	0.41	0.32	1.9
CD	0.44	0.32	2.3
DE	0.48	0.34	2.7
EF	0.48	0.33	2.9
FG	0.49	0.36	3.0
GH	0.49	0.35	3.1
HI	0.51	0.37	2.9
IJ	0.50	0.36	2.9
JK	0.52	0.37	2.8
KL	0.50	0.38	3.1
LM	0.49	0.35	2.8
MN	0.50	0.36	2.7
NO	0.47	0.34	2.5
OP	0.43	0.31	2.0
PQ	0.41	0.32	1.8
QR	0.39	0.30	1.6
RS	–	–	0.5

What is most nearly the total flow in the river?

(A) 1.3 m³/s
(B) 20 m³/s
(C) 40 m³/s
(D) 330 m³/s

Hint: Use section areas and average velocities.

PROBLEM 17

The 25 yr return period rainfall frequency-depth-duration curves for a coastal region are shown in the illustration. For a mean annual precipitation (P_{ma}) of 27 in, what is the rainfall intensity for a 2.5 hr storm?

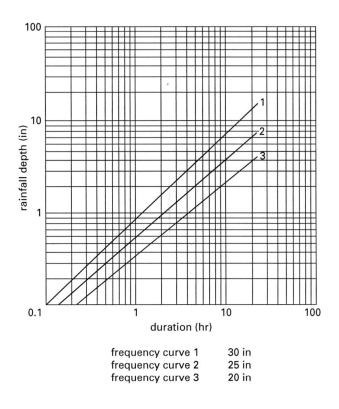

frequency curve 1	30 in
frequency curve 2	25 in
frequency curve 3	20 in

(A) 0.40 in/hr
(B) 0.60 in/hr
(C) 1.5 in/hr
(D) 11 in/hr

Hint: How do the units of the answer choices relate to the illustration?

PROBLEM 18

A 131 ac drainage area has the following characteristics and 10-year storm frequency-intensity-duration curve.

land use	area %
apartments	30
landscaped open space (park)	25
light industrial	45

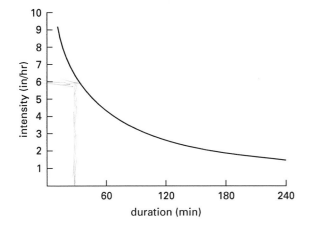

What is the peak runoff from the drainage area for the 30 min duration, 10-year storm?

(A) 24 ac-ft/hr
(B) 34 ac-ft/hr
(C) 72 ac-ft/hr
(D) 310 ac-ft/hr

Hint: Account for the variation in runoff coefficients for different land uses.

PROBLEM 19

A histogram of flood peak flows representing 112 events over a 94 year period is presented in the following illustration. If the largest 18 floods resulted in significant economic impact, what is the minimum peak flow of these floods?

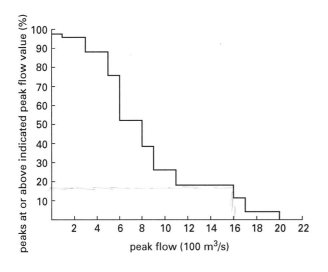

(A) 500 m³/s
(B) 1100 m³/s
(C) 1600 m³/s
(D) 16 000 m³/s

Hint: What format does the figure require of the data?

PROBLEM 20

Rainfall records for four precipitation stations are summarized in the given tables. Stations B, C, and D are those located in closest proximity to station A.

station	normal annual precipitation (cm)	annual precipitation for year indicated (cm)			
		1985	1986	1987	1989
A	39	32	37	42	41
B	31	27	30	37	34
C	42	34	39	45	46
D	37	34	32	–	39

station	normal annual precipitation (cm)	annual precipitation for year indicated (cm)			
		1990	1991	1992	1993
A	39	44	34	36	40
B	31	37	28	30	30
C	42	43	–	39	40
D	37	37	32	35	36

What is the estimated precipitation at station C for 1991?

(A) 31 cm
(B) 37 cm
(C) 40 cm
(D) 48 cm

Hint: The solution is not a simple average.

PROBLEM 21

The elemental hydrograph for a small urban drainage area is presented in the illustration.

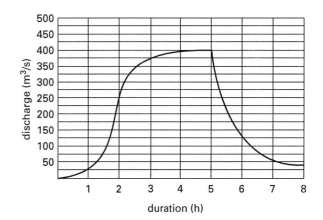

What is the volume of rainfall in surface detention?

(A) 1.2×10^6 m³
(B) 2.7×10^6 m³
(C) 4.5×10^6 m³
(D) 5.7×10^6 m³

Hint: What are the characteristics of an elemental hydrograph?

PROBLEM 22

Land use classification for a watershed in the western United States is summarized in the table. The longest flow path within the watershed to the storm sewer is 337 yd and the average ground slope over the watershed is 0.013.

land use	area (ac)
shingle roof	0.8
concrete surface	1.1
asphalt surface	2.6
poorly drained lawn	10

What is the flow to the storm sewer resulting from a 10 yr storm event?

(A) 1.7 ft^3/sec
(B) 6.3 ft^3/sec
(C) 9.2 ft^3/sec
(D) 11 ft^3/sec

Hint: The solution requires multiple steps to find time of concentration, intensity, and runoff.

PROBLEM 23

An average annual discharge record for a river covers a period of 83 yr. The average flow during this period was 1947 ft^3/sec with a standard deviation of 613 ft^3/sec. What is the magnitude of the 50-year flood?

(A) 1200 ft^3/sec
(B) 3700 ft^3/sec
(C) 4000 ft^3/sec
(D) 6100 ft^3/sec

Hint: The information given suggests an analysis method.

WASTEWATER TREATMENT

PROBLEM 24

Monitoring results from a sewer inflow and infiltration (I/I) evaluation are presented in the table.

section	total infiltration to section (m^3/d)	pipe diameter (mm)	pipe length (km)
1	2315	100	13.5
		200	6.8
		300	6.2
2	958	100	9.2
		200	7.1
		300	4.4
3	3996	100	24.9
		200	12.1
		300	11.9
4	1867	100	21.3
		200	11.0
		300	4.7

Which section of city sewer should receive first priority for rehabilitation?

(A) section 1
(B) section 2
(C) section 3
(D) section 4

Hint: Reduce the data for each section to a unit value that allows for direct comparison.

PROBLEM 25

What tank size is required to equalize the flow described by the following data?

time period (hr)	period average flow (10^6 gal/day)
0000–0400	1.39
0400–0800	3.21
0800–1200	4.05
1200–1600	2.63
1600–2000	3.91
2000–2400	1.98

(A) 5.0×10^5 gal
(B) 1.6×10^6 gal
(C) 2.1×10^6 gal
(D) 2.9×10^6 gal

Hint: Begin by preparing a table to construct a cumulative flow plot. The average flow is for each 4 hr period.

PROBLEM 26

An industrial plant currently pays a surcharge for discharge of untreated wastewater to a city sewer system. Plans to construct a wastewater pretreatment system have included flow monitoring with the following results.

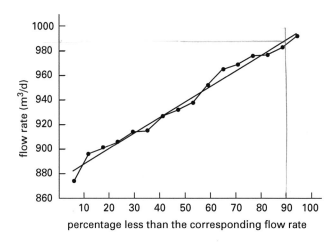

week	weekly average (m³/d)
1	976
2	901
3	938
4	965
5	977
6	992
7	874
8	914
9	906
10	927
11	969
12	932
13	983
14	896
15	915
16	952

What is the desired flow rate so that the flow does not exceed the capacity of the pretreatment system more than 10% of the time?

(A) 850 m³/d
(B) 890 m³/d
(C) 990 m³/d
(D) 1100 m³/d

Hint: Are the results more useful in the tabular or graphic form?

WATER TREATMENT

PROBLEM 27

A filter gallery of five multimedia sand filters treats 28,500 m³/d of water. Typically, each filter is backwashed twice during every 24 h period with backwashing occurring at 36 m³/m²·h for 25 min, followed by a conditioning period of 8 min. The filter loading rate is 225 m³/m²·d. Assume filtered water is used for backwashing. What is the approximate net daily production per filter?

(A) 4900 m³/d
(B) 5200 m³/d
(C) 5300 m³/d
(D) 5500 m³/d

Hint: What flow rates do the remaining filters see when one filter is backwashing?

PROBLEM 28

The Darcy friction factor for all pipes in the pipe network shown is 0.023 and all nodes are at equal elevation. Pipe lengths and nominal diameters are shown on the illustration.

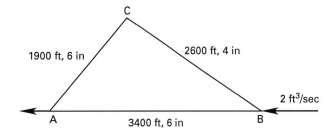

In this pipe network, what is the flow velocity in pipe BC?

(A) 0.56 ft/sec
(B) 2.9 ft/sec
(C) 6.5 ft/sec
(D) 17 ft/sec

Hint: The solution should converge quickly. If it does not, an error has been made.

PROBLEM 29

A planned community has a projected population of 3200 people. What is the maximum daily water demand during summer months, including fire demand?

(A) 790 gal/min
(B) 1900 gal/min
(C) 2200 gal/min
(D) 2600 gal/min

Hint: Are seasonal and peak multipliers applicable?

PROBLEM 30

The Safe Drinking Water Act (SDWA) includes all of the following EXCEPT

(A) regulation of all public drinking water systems in the U.S.
(B) regulation of both naturally occurring and man-made contaminants in drinking water.
(C) regulation of private drinking water wells serving less than 25 people.
(D) regulation of waste disposal through injection wells.

Hint: Consider the scope and purpose of the SDWA.

PROBLEM 31

Average diurnal water demand for a municipality is summarized in the table. A single water storage reservoir serves the municipality's population. Water is pumped to the reservoir continuously at a constant rate equal to the average daily demand. During what period of the day will the net water flow be into the reservoir?

time period (hr)	average demand (gal/min)
0000–0200	6900
0200–0400	6500
0400–0600	8100
0600–0800	12,100
0800–1000	14,300
1000–1200	15,600
1200–1400	13,800
1400–1600	12,500
1600–1800	9900
1800–2000	8900
2000–2200	8300
2200–2400	7200

(A) 6:00 a.m.–4:00 p.m.
(B) 4:00 p.m.–6:00 a.m.
(C) 4:20 p.m.–7:10 a.m.
(D) 4:40 p.m.–6:10 a.m.

Hint: Determine how the average daily flow relates to the average flow per period.

Depth Problems

HYDRAULICS—CLOSED CONDUIT

PROBLEM 32

An existing nominal 6 in steel water line is unable to meet the projected demand of 2.4 ft^3/sec for a growing residential development. A nominal 4 in steel line will be placed parallel to the existing line. Both pipes begin and end at the same point and are 1400 ft long. What will be the approximate flow in the 4 in pipe?

(A) 0.51 ft^3/sec
(B) 0.61 ft^3/sec
(C) 0.64 ft^3/sec
(D) 0.74 ft^3/sec

Hint: Assume an initial value for the Reynolds number.

PROBLEM 33

A submersible pump operated at 1750 rpm is needed to deliver 800 gal/min of water from a 350 ft deep well. At 93% efficiency, the average specific speed of the pump is 2300. How many stages are required for the pump?

(A) 1 stage
(B) 3 stages
(C) 6 stages
(D) 35 stages

Hint: Specific speed is derived from dynamic similarity.

PROBLEM 34

What is most nearly the maximum allowable suction head for a pump with a cavitation constant of 0.26 that is to operate under the following conditions?

elevation	1370 m above mean sea level
water temperature	13°C
total dynamic head	38 m

(A) 1.2 m
(B) 1.5 m
(C) 8.5 m
(D) 15 m

Hint: Define the cavitation constant.

PROBLEM 35

A new 5 km long pipeline connects two reservoirs. The water surface elevation in the upper reservoir is 1100 m and 835 m in the lower reservoir. The pipeline is 400 mm nominal diameter welded steel with a square mouth inlet and includes the following flanged fittings.

10 gate valves
19 standard radius 90° ells
37 standard radius 45° ells
8 straight tees

What is the maximum water flow rate between the reservoirs?

(A) 0.49 m^3/s
(B) 0.51 m^3/s
(C) 0.64 m^3/s
(D) 0.83 m^3/s

Hint: Use the energy equation and the Hazen-Williams equation.

PROBLEM 36

Consider the venturi meter shown. The upstream pipe diameter is 16 in, and the venturi throat diameter is 10 in. The fluid in the manometer is mercury, and the water and mercury temperatures are equal at 70°F.

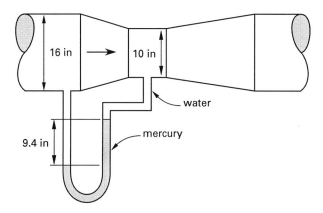

What is the flow rate of the given venturi meter?

(A) 5.9 ft^3/sec
(B) 15 ft^3/sec
(C) 23 ft^3/sec
(D) 38 ft^3/sec

Hint: Find an equation that relates the manometer fluid reading to the venturi meter velocity.

HYDRAULICS—OPEN CHANNEL

PROBLEM 37

A square baffled outlet is used to dissipate energy at the bottom of a rectangular chute with a 3 ft base width. The head and flow are 42 ft and 200 ft^3/sec, respectively. The Froude number and the basin width to channel outlet depth ratio are related by $w_b/D_c = 0.875\,\text{Fr} + 2.7$. What is the minimum required width of the baffled outlet basin?

(A) 1.3 ft
(B) 2.6 ft
(C) 12 ft
(D) 26 ft

Hint: Do not confuse the outlet width with the basin width.

PROBLEM 38

Water flows at a depth of 1.5 m and a velocity of 2.5 m/s in an 8 m wide rectangular open channel. The channel transitions to a circular culvert to cross 50 m under a highway. Both the channel and the culvert are constructed of concrete at a constant slope of 0.002 m/m. What is the required culvert diameter if the culvert is to be constructed of standard concrete pipe and is to flow half full?

(A) two pipes, 9 ft diameter
(B) three pipes, 9 ft diameter
(C) four pipes, 9 ft diameter
(D) eight pipes, 9 ft diameter

Hint: Does the continuity equation apply?

PROBLEM 39

The water height above the crest of a dam spillway is 2 m and the spillway crest is 21 m above the bottom of the reservoir.

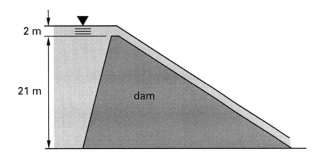

What value most nearly represents the force of the water exerted on the dam?

(A) 17 kN/m
(B) 340 kN/m
(C) 2300 kN/m
(D) 2500 kN/m

Hint: Be careful of energy and depth relationships.

PROBLEM 40

An ordinary firm loam-lined ditch is used to convey clear irrigation water from a reservoir to fields below. The first 24 m section of the ditch follows a slope of 5%, with discharge occurring over a weir that can be raised or lowered to control flow. The ditch has a 90° V-shaped cross section and 20 cm diameter logs are secured in the ditch channel perpendicular to the flow direction at 4.0 m spacing along the steep section. The scour velocity for clear water flowing over ordinary firm loam is 0.76 m/s. What is the approximate maximum flow in the ditch if the scour velocity should not be exceeded?

(A) 0.0014 m^3/s
(B) 0.033 m^3/s
(C) 0.064 m^3/s
(D) 0.28 m^3/s

Hint: When does the maximum water velocity occur?

PROBLEM 41

Construction equipment has punctured a pipeline that runs parallel to a 6 ft wide concrete-lined drainage ditch. The puncture is 1.8 miles upstream of where the ditch discharges into a saltwater marsh. 200 gal of a chemical were released from the puncture. The ditch normally flows at 3 ft/sec with a water depth of 2 ft. The average slope of the ditch channel is 0.01. The specific gravity of the chemical is 0.9 and the ditch dispersion coefficient is 0.3 day^{-1}. What will be the maximum concentration of the chemical when it reaches the marsh?

(A) 6000 mg/L
(B) 6700 mg/L
(C) 6800 mg/L
(D) 8500 mg/L

Hint: Look for references regarding an instantaneous release into a river, channel, or estuary.

HYDROLOGY

PROBLEM 42

A commercial sod farm maintains 500 ha in current production year-round. The sod requires about 260 cm of water annually, and is applied three times weekly to subdivided plots by a fixed sprinkler irrigation system at an application rate of 2 cm/h. What is the maximum subdivided plot size if irrigation at the sod farm is to occur for 12 h/d and 7 d/wk?

(A) 5 ha
(B) 15 ha
(C) 21 ha
(D) 45 ha

Hint: Use time ratios.

PROBLEM 43

The water surface in a reservoir is maintained at the spillway elevation by sluiceways except during periods of heavy rainfall. During dry weather conditions the average flow into the reservoir is 1100 ft^3/sec, but during the 10-year storm, inflow increases to 1550 ft^3/sec after 8 hr and the water level rises 6 ft above the spillway crest. The reservoir storage capacity when the pool elevation is at the spillway crest is 160,000 ac-ft and the spillway is 30 ft wide. By approximately how much is the reservoir storage increased during the 10 yr storm?

(A) 200 ac-ft
(B) 430 ac-ft
(C) 560 ac-ft
(D) 160,000 ac-ft

Hint: Use critical depth and energy.

PROBLEM 44

A rainfall frequency-depth-duration curve for a watershed is shown in the following illustration. The value for the Steel formula constant K is 180 in-min/hr and b is 25 min.

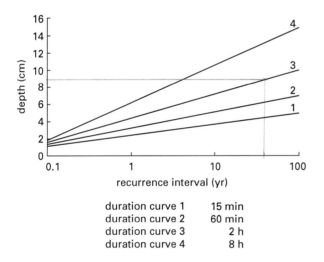

duration curve 1	15 min
duration curve 2	60 min
duration curve 3	2 h
duration curve 4	8 h

What is the time of concentration for the 2 h duration 50 yr storm?

(A) 15 min
(B) 26 min
(C) 42 min
(D) 75 min

Hint: How does the Steel formula apply?

PROBLEM 45

City parking lots A and B slope to join at a common gutter that discharges to a storm sewer inlet located at one end of the lots, as shown. Both lots are 100 ft wide, with lot A being 150 ft deep and lot B being 90 ft deep. Lot A slopes at 0.001 and lot B slopes at 0.0006. Both lots are newly paved with smooth asphalt.

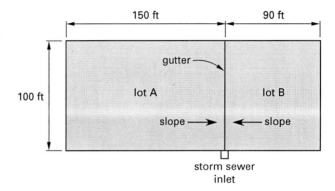

For a rainfall intensity of 0.89 in/hr over a 1 hr period, what is the time of concentration for flow to the gutter?

(A) 0.14 min
(B) 18 min
(C) 28 min
(D) 36 min

Hint: The small drainage area and the form of the rainfall data dictate selection of the appropriate equation.

PROBLEM 46

Consider the unit hydrograph developed for a 3 h storm and a 420 km^2 drainage area, as shown.

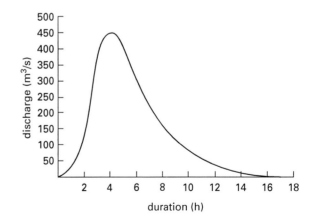

What is the peak flow of three successive 3 h periods of rainfall that produce 1.6 cm, 3.1 cm, and 2.7 cm of runoff?

(A) 1400 m^3/s
(B) 2100 m^3/s
(C) 2900 m^3/s
(D) 3300 m^3/s

Hint: Add the hydrographs.

PROBLEM 47

The illustration shown presents a moisture-tension curve for a soil.

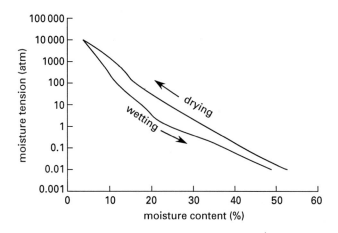

What is the moisture content of the soil that corresponds to the wilting point?

(A) 17%
(B) 20%
(C) 23%
(D) 32%

Hint: How is the wilting point defined?

PROBLEM 48

The feasibility of a dam for impounding irrigation water in the southwestern United States is to be determined, in part, by the volume of water lost from the resulting reservoir to evaporation. The water temperature will average 18°C and the air temperature within 2 m of the water surface over the reservoir will average 24°C. The air temperature at the water surface is the same as the water temperature. The relative humidity is 16%. The wind speed over the reservoir at 4 m above the water surface will average 3.6 m/s. What will be the approximate daily water loss from the reservoir to evaporation?

(A) 0.17 mm/d
(B) 4.0 mm/d
(C) 5.2 mm/d
(D) 7.0 mm/d

Hint: What estimation method is suggested by the data provided? Be careful when defining vapor pressure.

PROBLEM 49

The average annual sediment load to a reservoir is presented in the following illustration. The sediment is composed of 64% clay and 36% silt and will always be submerged. The original volume of the reservoir was 3600 ac-ft.

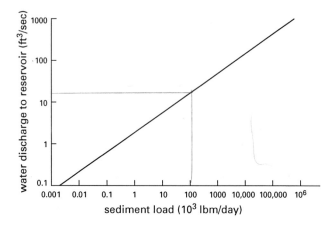

For an average annual stream flow of 20 ft³/sec, what is most nearly the useful life of the reservoir if its minimum useful volume is 75% of its original volume?

(A) 14 yr
(B) 59 yr
(C) 240 yr
(D) 960 yr

Hint: Be careful of how the specific weights of the sediments are determined.

PROBLEM 50

The inflow, outflow, and water surface elevation for a deep reservoir with an uncontrolled discharge are summarized in the table.

inflow (m³/s)	outflow (m³/s)	water surface elevation (m)
0.67	0.67	371.1
1.2	0.70	373.2
2.8	2.9	379.1
2.5	3.8	377.8

The elevation of the spillway crest is 371 m. The equation to estimate the reservoir volume in m³ is

$$s = 9.4(\text{elevation, m})^2 + 3854(\text{elevation, m}) + 14\,470$$

How much time is required for the reservoir to reach its peak storage, coinciding with a maximum water surface elevation of 377.8 m?

(A) 3.9 h
(B) 7.6 h
(C) 30 h
(D) 130 h

Hint: Use mass balance or find an equation for routing through an uncontrolled reservoir.

PROBLEM 51

Frequency-intensity-duration curves and a sketch of a watershed that will be developed into a theme park are presented in the illustrations. All land within the watershed will be paved with an average ground slope of 1.2%.

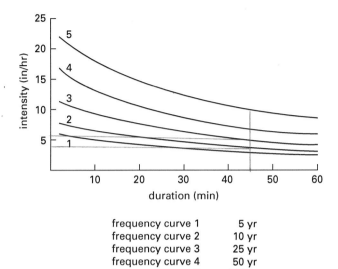

frequency curve 1	5 yr
frequency curve 2	10 yr
frequency curve 3	25 yr
frequency curve 4	50 yr
frequency curve 5	100 yr

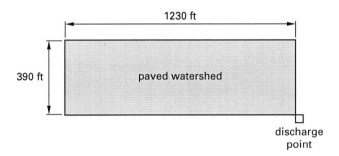

What is the change in the time of concentration from the 10-year to the 25-year storm for a storm duration of 45 min?

(A) 0.70 min
(B) 2.3 min
(C) 2.8 min
(D) 30 min

Hint: Find an equation appropriate for paved surfaces that includes rainfall intensity as a parameter.

PROBLEM 52

A reservoir is proposed for storing irrigation water. To meet irrigation demand, the required minimum annual yield of the reservoir must be 90,000 ac-ft. The reservoir will receive water through inflow from several small streams. The mass inflow from the streams over the preceding 3 yr period is presented in the figure.

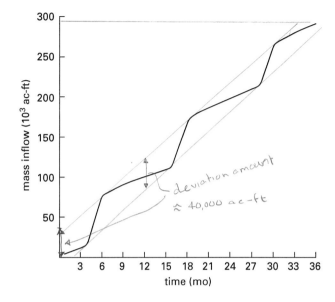

What reservoir capacity is required to meet the minimum yield?

(A) 40,000 ac-ft
(B) 78,000 ac-ft
(C) 90,000 ac-ft
(D) 270,000 ac-ft

Hint: What is the relationship between the information provided in the figure and the average annual demand?

GROUNDWATER AND WELL FIELDS

PROBLEM 53

Using the given information, what is the hydraulic conductivity of the aquifer for nonaqueous phase liquid (NAPL)?

hydraulic conductivity (water)	2.0×10^{-4} cm/s
aquifer temperature	10°C
density NAPL at 10°C	0.92 g/cm^3
dynamic viscosity NAPL at 10°C	0.066 g/cm·s

(A) 1.9×10^{-7} cm/s
(B) 3.6×10^{-5} cm/s
(C) 4.0×10^{-5} cm/s
(D) 1.8×10^{-4} cm/s

Hint: Distinguish between hydraulic conductivity and intrinsic permeability and review how they are related.

PROBLEM 54

What is the approximate solute actual velocity for the site depicted in the illustration if the hydraulic conductivity is 0.83 ft/day, the soil porosity is 0.37, and the retardation factor is 1.94?

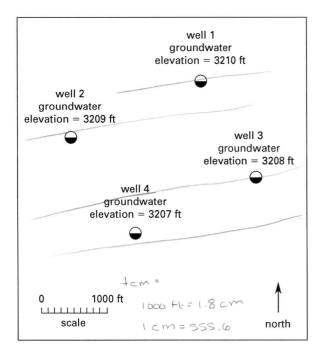

(A) 0.00056 ft/day
(B) 0.0015 ft/day
(C) 0.0021 ft/day
(D) 0.0029 ft/day

Hint: Be careful to distinguish between actual and Darcy velocity. Will the retardation factor increase or decrease the solute velocity compared to the groundwater velocity?

PROBLEM 55

An abandoned industrial site with soil and groundwater contaminated by an organic solvent has the following characteristics.

gradient	0.00063
effective porosity	0.38
intrinsic permeability	1.1×10^{-5} mm^2
soil total organic carbon	485 mg/kg
soil bulk density	1.8 g/cm^3
soil-water partition coefficient	173 mL/g
temperature	10°C

What is the velocity of the dissolved organic solvent?

(A) 2.2×10^{-11} m/d
(B) 5.5×10^{-5} m/d
(C) 0.0084 m/d
(D) 0.012 m/d

Hint: Be careful to use correct definitions for the terms used in the equations.

PROBLEM 56

A small pest-control business routinely discharges pesticide-contaminated water to a drainage ditch when washing their equipment. The ditch infiltrates to a shallow aquifer with a bulk groundwater velocity of 172 cm/d. The concentration of the pesticide in the aquifer just below the ditch is 0.182 mg/L. How much time will be needed for the pesticide to reach a drinking water well located in a direct line 1600 m downgradient of the source at its maximum contaminant level (MCL) of 1.0 μg/L?

(A) 0.8 d
(B) 120 d
(C) 860 d
(D) 5300 d

Hint: Make simplifying assumptions and choose an equation for the appropriate boundary conditions.

PROBLEM 57

A saturated soil profile is characterized in the table. The groundwater elevation difference between two wells, one that screens layer 1 and the other that screens layer 4, is 14 cm.

layer	thickness (cm)	hydraulic conductivity (cm/s)
1	130	0.0090
2	180	0.017
3	270	0.036
4	65	0.011
5	110	0.020

What is the vertical groundwater flow between the screened layers?

(A) 2.18×10^{-5} m^3/s for 1 m^2 of aquifer area
(B) 2.35×10^{-5} m^3/s for 1 m^2 of aquifer area
(C) 2.41×10^{-5} m^3/s for 1 m^2 of aquifer area
(D) 2.56×10^{-5} m^3/s for 1 m^2 of aquifer area

Hint: How is the overall hydraulic conductivity calculated?

PROBLEM 58

A well completely penetrates a confined aquifer under conditions shown in the following illustration. The hydraulic conductivity of the aquifer is 7.3 ft/day.

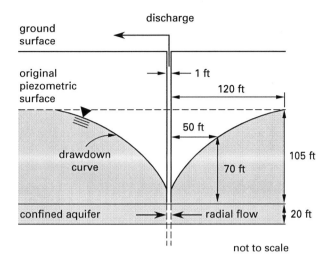

What is the flow from the well?

(A) 0.029 ft^3/sec
(B) 0.16 ft^3/sec
(C) 0.43 ft^3/sec
(D) 1.9 ft^3/sec

Hint: Discharge flow equations are available for confined and unconfined aquifers. Use the illustration to define terms.

PROBLEM 59

An agricultural drain field is depicted in the illustration.

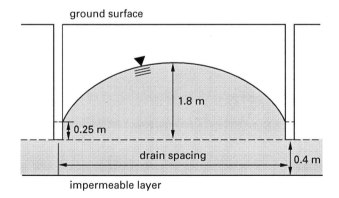

For a soil hydraulic conductivity of 0.018 cm/s and an average infiltration rate of 0.0014 cm/s, what is the required drain spacing?

(A) 1.2 m
(B) 15 m
(C) 140 m
(D) 230 m

Hint: This is a classic drain spacing problem. Look for a drain spacing equation and be careful that the terms match those given in the illustration.

WASTEWATER TREATMENT

PROBLEM 60

Complete mix activated sludge has been selected for treatment of a wastewater.

flow rate	25×10^6 gal/day
reactor volume	5×10^6 gal
influent biochemical oxygen demand (BOD) concentration	224 mg/L
effluent BOD concentration	20 mg/L
reactor mixed liquor suspended solids concentration	3500 mg/L
recirculated solids concentration	12 000 mg/L
mean cell residence time	10 day
yield coefficient	0.5 g/g
endogenous decay rate constant	0.05/day

What is most nearly the recirculated solids flow rate required to maintain the mean cell residence time?

(A) 1.0×10^7 gal/day
(B) 1.8×10^7 gal/day
(C) 2.5×10^7 gal/day
(D) 6.0×10^7 gal/day

Hint: Begin by deriving an equation for the recirculated solids flow rate. Consider a mass balance around the clarifier.

PROBLEM 61

A wastewater treatment process wastes sludge at 50,000 gal/day. The wasted sludge contains 1.2% solids. What volume reduction can be realized by thickening and dewatering the sludge to 24% solids?

(A) 2500 gal/day
(B) 2600 gal/day
(C) 39,000 gal/day
(D) 48,000 gal/day

Hint: This problem can be solved by simple ratios and differences.

PROBLEM 62

An electronics component manufacturer generates wastewater at 135 gal/min containing 12 mg/L of methylene chloride. A concentration of 100 µg/L was set as the treatment criteria for reuse of the water. The average water temperature is 25°C. The mass transfer coefficient using the selected packing material is 0.023.

What is most nearly the minimum required air flow rate if air stripping is the selected removal process? Assume a stripping factor of 3.5.

(A) 4.0 ft³/min
(B) 63 ft³/min
(C) 490 ft³/min
(D) 900 ft³/min

Hint: Units of Henry's constant are important to obtain the correct solution. Select these so the air-to-water ratio is unitless.

PROBLEM 63

Settling removal efficiency curves of a settling basin designed for a Type II suspension are presented in the illustration.

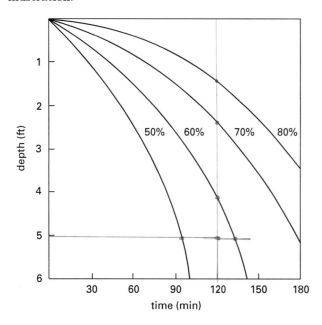

What is the removal efficiency at a depth of 5 ft and settling time of 120 min?

(A) 43%
(B) 57%
(C) 66%
(D) 71%

Hint: Incremental efficiency occurs at depth intervals.

PROBLEM 64

An anaerobic lagoon will pretreat a slaughterhouse wastewater that will be discharged to an existing facultative lagoon. Influent flow is 2.6×10^5 gal/day with a total biochemical oxygen demand (BOD) of 14 000 mg/L. The average waste temperature during winter months is 21°C and during summer months is 27°C. The winter loading rate is 12 lbm BOD/10³ ft³-day and the summer loading rate is 18 lbm BOD/10³ ft³-day. Site conditions limit the lagoon depth to 10 ft. What is the required total surface area of the anaerobic lagoon?

(A) 3.9 ac
(B) 4.6 ac
(C) 5.0 ac
(D) 5.8 ac

Hint: How do loading rates influence design?

PROBLEM 65

The following are selected characteristics of an activated sludge process bioreactor.

influent flow rate	27 000 m³/d
influent biochemical oxygen demand (BOD)	281 mg/L
effluent BOD	20 mg/L
yield coefficient	0.53 g/g
endogenous decay rate constant	0.05 d⁻¹
mean cell residence time	8 d

What is the approximate daily mass of biomass produced in the bioreactor?

(A) 2700 kg/d
(B) 3700 kg/d
(C) 7000 kg/d
(D) 16 000 kg/d

Hint: Review the definition of the observed yield coefficient.

PROBLEM 66

What is most nearly the required media total surface area for a rotating biological contactor (RBC) process selected to treat the following wastewater?

flow rate	250,000 gal/day
influent total biochemical oxygen demand (BOD)	174 mg/L
effluent soluble BOD	30 mg/L

(A) 130,000 ft²
(B) 150,000 ft²
(C) 180,000 ft²
(D) 360,000 ft²

Hint: The solution to this problem is based on loading rates.

PROBLEM 67

Constructed wetlands for wastewater treatment are characterized as submerged flow (SF) and free water surface (FWS). In which of the following scenarios would an FWS wetland be preferred over an SF wetland?

(A) An SF wetland provides improved odor and vector control over an FWS wetland.
(B) An SF wetland provides improved suspended solids removal over an FWS wetland.
(C) An SF wetland provides improved ammonia removal compared to an FWS wetland.
(D) An SF wetland is less susceptible to temperature and other climate extremes than are FWS wetlands.

Hint: Review the design and operating conditions of CWWT.

PROBLEM 68

Bench scale bioreactors operated to model activated sludge treatment of wastewater produced the following results. The reactors are completely mixed without solids recycle.

reactor	influent biochemical oxygen demand (mg/L)	effluent biochemical oxygen demand (mg/L)	hydraulic residence time (d)	mixed liquor suspended solids (mg/L)
1	210	10	3.45	118
2	210	16	1.92	135
3	210	23	1.45	136
4	210	36	1.10	132

What is the value of the endogenous decay-rate coefficient?

(A) -1.2 d^{-1}
(B) -0.10 d^{-1}
(C) 0.14 d^{-1}
(D) 0.87 d^{-1}

Hint: Find an equation that includes the endogenous decay rate coefficient with the parameters given in the table.

PROBLEM 69

What is most nearly the daily mass of oxygen required for aeration if nitrification is to occur for the following wastewater?

flow rate	5.0×10^6 gal/day
influent ammonia concentration	63 mg/L
effluent ammonia concentration	10 mg/L
influent five-day biochemical oxygen demand (BOD$_5$) concentration	356 mg/L
effluent BOD$_5$ concentration	30 mg/L
ratio of BOD$_5$ to ultimate BOD (BOD$_u$)	1:1.52
daily mass of biomass wasted	2800 lbm/day

(A) 17,000 lbm/day
(B) 20,000 lbm/day
(C) 27,000 lbm/day
(D) 31,000 lbm/day

Hint: Consider which conditions contribute to oxygen demand.

PROBLEM 70

What is the corrected maximum growth rate for the nitrifying system with the following characteristics?

pH	6.4
temperature	17°C
dissolved oxygen concentration	7.2 mg/L
yield coefficient	0.23 g/g
endogenous decay rate constant	0.05 d^{-1} at 17°C
growth rate constant	2.1 d^{-1} at 17°C

(A) 0.046 d^{-1}
(B) 0.17 d^{-1}
(C) 0.48 d^{-1}
(D) 0.72 d^{-1}

Hint: One or more typical parameter values are assumed.

PROBLEM 71

A chemical formula typically used to represent biomass is given as $C_5H_7O_2N$. What is most nearly the daily nitrogen requirement if the wastewater contains 310 mg/L acetic acid and the flow rate is 12 000 m^3/d?

(A) 140 kg/d
(B) 170 kg/d
(C) 350 kg/d
(D) 430 kg/d

Hint: Begin with the mole ratio.

PROBLEM 72

An industrial wastewater discharge of 2.5×10^5 gal/day contains n-butylphthalate at an average concentration of 148 mg/L. The required removal efficiency is 99%. The isotherm constants for n-butylphthalate are an intercept of 220 mg/g and slope of 0.45. Which of the following options is the smallest granular activated carbon (GAC) adsorber that can provide a minimum carbon change-out period of 14 days?

(A) 2000 lbm
(B) 4000 lbm
(C) 10,000 lbm
(D) 20,000 lbm

Hint: What do the intercept and slope units suggest for an isotherm equation?

PROBLEM 73

A hazardous waste generated at 50,000 gal/day and with a pH of 1.6 is treated by simple neutralization using 0.005 N sodium hydroxide. What is the daily sodium hydroxide feed rate?

(A) 0.015 mL/d
(B) 3.8×10^3 mL/d
(C) 1.5×10^5 mL/d
(D) 9.5×10^{12} mL/d

Hint: Consider the relationship between pH and pOH.

PROBLEM 74

A plating line uses ion exchange to recover chromic acid, as CrO_3, from a 15% solution generated at 1000 m³/d. The exchange resin deteriorates rapidly at CrO_3 concentrations greater than 11%. The recovered solution generated at 350 m³/d contains 42% CrO_3, and to meet reuse specifications, the recovered CrO_3 must be diluted to a 3.5 molar concentration. The ion exchange system requires what daily volume of dilution water?

(A) 430 m³/d
(B) 1300 m³/d
(C) 1800 m³/d
(D) 2200 m³/d

Hint: Find the dilution ratio for each step.

PROBLEM 75

The biological degradation of an organic chemical in groundwater is represented by the following equation.

$$26CH_3COO^- + 6NH_4^+ + 21O_2$$
$$\rightarrow 6C_5H_7O_2N + 18H_2O + 2CO_2 + 20HCO_3^-$$

The groundwater contains ammonia from septic tank leachate at 9.7 mg/L and the contaminant is present at 818 mg/L. If no nitrogen augmentation is included, by approximately how much will the contaminant concentration be reduced through biodegradation?

(A) 7.4 mg/L
(B) 58 mg/L
(C) 140 mg/L
(D) 810 mg/L

Hint: Start with ammonia.

WATER QUALITY

PROBLEM 76

Which of the following characteristics are included in the minimum national standards for secondary wastewater treatment under the Clean Water Act (CWA)?

(A) suspended solids and five day biochemical oxygen demand
(B) suspended solids and disinfection byproducts
(C) five day biochemical oxygen demand and dissolved solids
(D) disinfection byproducts and dissolved solids

Hint: The CWA addresses discharges to receiving waters.

PROBLEM 77

Among the common types of pathogenic organisms listed, which of them is infrequently looked for in routine analysis, generally less susceptible to chlorination, and targeted for removal by filtration?

(A) protozoa
(B) viruses and bacteria
(C) viruses and helminths
(D) protozoa and helminths

Hint: Generally, as microorganisms increase in size and complexity, especially in reproductive method, they are more likely to be resistant to chlorination and more likely to be removed by filtration.

PROBLEM 78

What is most nearly the five day biochemical oxygen demand (BOD_5) at 15°C of the water represented by the tabulated data? The samples were incubated for five days at 20°C. The rate coefficient is 0.17 d^{-1} (base 10).

bottle	sample volume (mL)	dissolved oxygen at $t = 0$ d (mg/L)	dissolved oxygen at $t = 5$ d (mg/L)
1	20	9.0	1.0
2	10	9.1	2.9
3	5	9.1	6.1
4	2	9.2	8.1

(A) 140 mg/L
(B) 150 mg/L
(C) 170 mg/L
(D) 180 mg/L

Hint: Review the data to see if the problem can be simplified by disregarding some of the bottles. Remember, you have to calculate ultimate BOD before applying the temperature correction.

PROBLEM 79

500 mL of wastewater with an initial pH of 7.3 is titrated with 0.03 N H_2SO_4. The 4.5 endpoint pH is reached when 14.5 mL of acid have been added. What is the concentration of the bicarbonate alkalinity?

(A) 22 mg/L as $CaCO_3$
(B) 29 mg/L as $CaCO_3$
(C) 44 mg/L as $CaCO_3$
(D) 72 mg/L as $CaCO_3$

Hint: The concentration of the acid used determines the amount of alkalinity neutralized and the initial pH defines what alkalinity species are dominant.

PROBLEM 80

The dissolved oxygen concentration is 9.3 mg/L and the ultimate biochemical oxygen demand concentration is 9.8 mg/L where a wastewater discharges to a stream. The stream flows at 0.3 ft/sec, and the stream water temperature is 8.6°C with reoxygenation and deoxygenation rate constants equal at 0.5 day^{-1} and 0.4 day^{-1}, respectively. Approximately how far downstream from the discharge point should the monitoring station be located to detect the maximum oxygen deficit caused by the discharge? The rate constants are given for the natural log (ln, e).

(A) 0.079 mi
(B) 2.4 mi
(C) 7.9 mi
(D) 180 mi

Hint: This is a critical-time problem.

PROBLEM 81

Arsenic, cadmium, and fluoride have been detected in the soil of a city park at 1.3 ppb, 0.96 ppb, and 0.42 ppb, respectively. The oral route reference dose for arsenic is 0.0003 mg/kg·d, for cadmium is 0.0005 mg/kg·d, and for fluoride is 0.0003 mg/kg·d. What is the hazard index if the exposed population is children who may ingest the soil?

(A) 1.1×10^{-12}
(B) 8.3×10^{-6}
(C) 0.00010
(D) 0.51

Hint: The hazard index is for noncarcinogenic exposure.

PROBLEM 82

Which one of the following would likely NOT be effective for controlling algae in wastewater effluents?

(A) aeration
(B) microscreening
(C) nitrification/denitrification
(D) chlorination

Hint: Consider the treatment function of each choice.

PROBLEM 83

A stream segment below a stormwater discharge exhibits the following characteristics.

saturated dissolved oxygen concentration	10.9 mg/L
mixed ultimate biochemical oxygen demand (BOD_u) at the discharge	7.2 mg/L
dissolved oxygen deficit at the discharge point	3.2 mg/L
reaeration constant (ln, e)	0.07 d^{-1}
deoxygenation constant (ln, e)	0.04 d^{-1}

Which oxygen sag curve represents the stream's dissolved oxygen profile below the discharge?

(A)

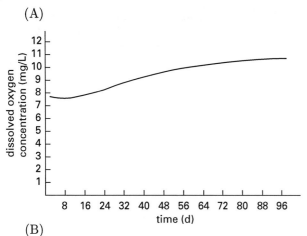

(B)

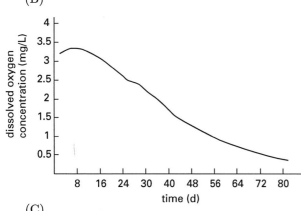

(C)

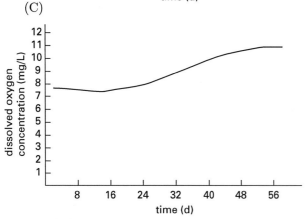

(D)

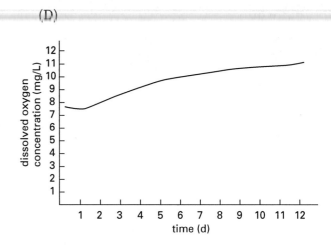

Hint: Oxygen sag curves are characterized by a typical profile.

PROBLEM 84

Effluent limits for a proposed wastewater discharge to a river allow 65% recovery of the river's dissolved oxygen (DO) concentration within three days. The average river flow is 37 ft^3/sec with an ultimate biochemical oxygen demand (BOD_u) of 6 mg/L upstream of the discharge. The illustration shows the dissolved oxygen profile for the river as a function of mixed-flow BOD_u (L_o). The projected discharge flow is 12×10^6 gal/day (12 MGD).

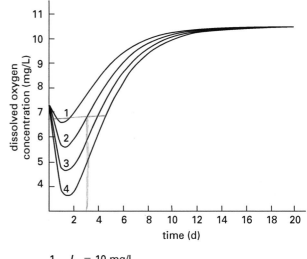

1. L_o = 10 mg/L
2. L_o = 14 mg/L
3. L_o = 18 mg/L
4. L_o = 22 mg/L

What is the maximum allowable BOD_u of the discharge?

(A) 14 mg/L
(B) 30 mg/L
(C) 39 mg/L
(D) 54 mg/L

Hint: Start by examining the illustration. What information does it provide?

PROBLEM 85

The illustration represents a chromatograph for gasoline sampled from an underground storage tank.

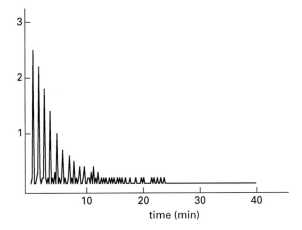

Assume the underground storage tank leaked some of the gasoline to groundwater. Which of the following chromatographs would most likely represent the same gasoline as sampled from a monitoring well?

(A)

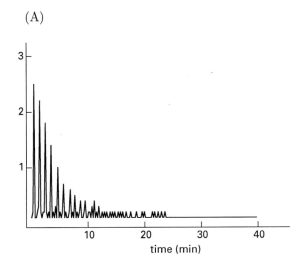

(B)

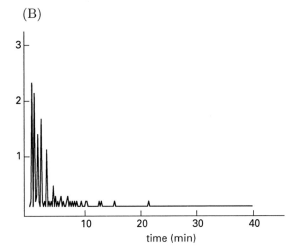

(C)

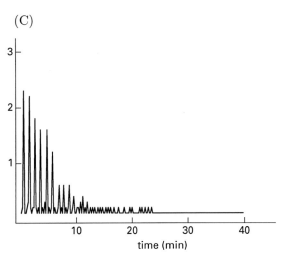

(D)

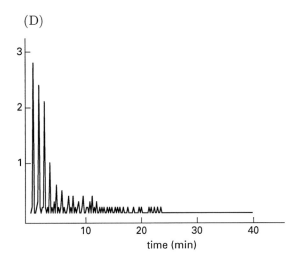

Hint: How would weathering influence the chromatograph?

WATER TREATMENT

PROBLEM 86

A water supply contains a total hardness of 382 mg/L as $CaCO_3$. The water treatment plant uses ion exchange to provide water to its customers with a total hardness to 100 mg/L as $CaCO_3$ at a flow rate of 130 000 m³/d. Reactor vessels with a bed volume of 4 m³ are available and the resin capacity is 95 kg/m³. What is the total number of reactor vessels required if regeneration occurs once daily?

(A) 34
(B) 71
(C) 97
(D) 130

Hint: How is the desired hardness retained in the water supplied to the customers?

PROBLEM 87

A given tank-impellor flash mixer has a design velocity gradient of 700 sec⁻¹, a residence time of 2 min, and a motor efficiency of 88%. If the tank-impellor flash mixer is sized to treat 5×10^6 gal/day at a water temperature of 60°F, what is the required motor size?

(A) 15 hp
(B) 20 hp
(C) 22 hp
(D) 25 hp

Hint: How do efficiency and standard motor size influence the solution?

PROBLEM 88

A flocculation basin consists of two sections each with an equal length and depth of 4.0 m and a width of 8 m. The average velocity gradient is 45 s⁻¹ and the paddle speed in both sections is 3 rpm with flat paddles. The paddles are horizontally mounted with the axis perpendicular to flow. The water temperature is 20°C. What is the required paddle area in the first section of the flocculation basin?

(A) 0.72 m²
(B) 1.9 m²
(C) 2.8 m²
(D) 4.5 m²

Hint: Consider paddle-wall clearance and paddle slip.

PROBLEM 89

A sedimentation basin with a settling zone surface area of 5700 ft² accepts a flow of 2.7×10^6 gal/day. For a particle-settling velocity of 0.008 in/sec, what is the approximate removal efficiency of the sedimentation basin?

(A) 1.5%
(B) 38%
(C) 46%
(D) 92%

Hint: Is there a relationship between overflow rate and particle-settling velocity?

PROBLEM 90

A reverse osmosis (RO) system is required to treat a drinking water source that is subject to saltwater intrusion. The water and RO system have the following characteristics.

desired fresh water flow rate	30 000 m³/d
permeate recovery	77%
membrane flux rate	0.93 m³/m²·d
membrane packing density	800 m²/m³
membrane module volume	0.028 m³
pressure vessel capacity	12 modules

How many pressure vessels are required to treat the water?

(A) 92
(B) 120
(C) 160
(D) 1900

Hint: Determine the flow rate that requires treatment.

PROBLEM 91

What is the head loss through a clean single media sand filter defined by the following characteristics?

water temperature	20°C
clean filtering velocity	0.003 m³/m²·s
media bed depth	0.75 m
media mesh	12 × 16

(A) 2.9 cm
(B) 7.5 cm
(C) 19 cm
(D) 25 cm

Hint: Start with the Reynolds number, but be careful of how the particle diameter is determined.

PROBLEM 92

For a flow rate of 10 000 m³/d, what is the approximate overflow rate for a settling basin to achieve 80% efficiency? Results from a settling column test for the Type I suspension are shown in the illustration.

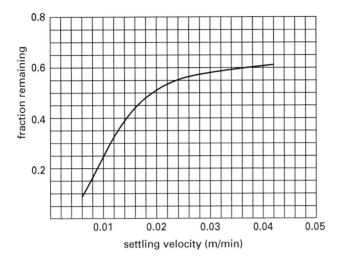

(A) 0.0085 m/min
(B) 0.011 m/min
(C) 0.015 m/min
(D) 0.02 m/min

Hint: Integrate the curve.

PROBLEM 93

An analysis of water requiring softening treatment is as follows.

ion	concentration (mg/L)
Ca^{2+}	187
Mg^{2+}	49
HCO_3^-	618

The flow rate of the water is 30 000 m³/d and the total hardness is to be removed to 100 mg/L as $CaCO_3$. the caustic soda is available at 85% purity. A tonne is 1000 kg. What is the monthly caustic soda mass required for softening the described water?

(A) 460 tonne/mo
(B) 560 tonne/mo
(C) 610 tonne/mo
(D) 650 tonne/mo

Hint: Use molar concentrations.

PROBLEM 94

A 250 mL water sample with an initial pH of 9.7 is titrated with 0.03 N H_2SO_4. A pH of 8.3 is reached after 6 mL of acid are added and a pH of 4.5 is reached after another 12 mL of acid are added. Which alkalinity species dominates and what is its concentration?

(A) carbonate dominates at 43 mg/L as $CaCO_3$
(B) carbonate dominates at 67 mg/L as $CaCO_3$
(C) carbonate dominates at 72 mg/L as $CaCO_3$
(D) bicarbonate dominates at 72 mg/L as $CaCO_3$

Hint: How does the initial pH influence the distribution of the alkalinity species?

PROBLEM 95

Disinfection of a wastewater using aqueous chlorine at a pH of 8.5 and a temperature of 21°C requires 23 min to effect the desired percentage kill. How much time is required if the wastewater temperature is 17°C?

(A) 6.4 min
(B) 20 min
(C) 27 min
(D) 32 min

Hint: Use activation energy to correct for temperature.

PROBLEM 96

Reverse osmosis having the following characteristics was selected to treat a drinking water source with high total dissolved solids.

desired freshwater flow rate	16 000 m³/d
permeate recovery	80%
salt rejection	92%
membrane flux rate	0.83 m³/m²·d
membrane packing density	800 m²/m³
membrane module volume	0.03 m³
pressure vessel capacity	10 modules

How many pressure vessels are required to treat the water?

(A) 63
(B) 80
(C) 87
(D) 100

Hint: Look for the membrane volume.

ENGINEERING ECONOMICS

PROBLEM 97

A water treatment plant serves a population of 65,000 people. Treatment costs are $0.12/(10)^3$ gal with 20% attributable to electrical power. Assuming continuous operation, what is the approximate annual electric bill for the plant?

(A) $58,000/yr
(B) $67,000/yr
(C) $94,000/yr
(D) $2,300,000/yr

Hint: What is the typical average annual daily per capita water demand?

PROBLEM 98

A municipality with a population of 215,000 is under state mandate to recycle 25% of the solid waste generated by its citizens. The remaining 75% will be landfilled. The per capita waste generation rate is 4.6 lbm/day. The landfilled waste in-place maximum compacted density is 50 lbm/ft^3, and the soil-cover-to-compacted-waste ratio is 1:4.5 by volume. The landfill covers a rectangular area 1200 ft by 1600 ft. The maximum landfill height cannot exceed 80 feet with 1:1 side slopes. What is the operating life of the landfill?

(A) 5 yr
(B) 14 yr
(C) 21 yr
(D) 64 yr

Hint: Be careful applying the cover-to-fill ratio and distinguishing between the waste landfilled and the waste recycled.

PROBLEM 99

A city anticipates the need for a solid waste transfer station within the coming few years. The amortized capital and operating cost of the transfer station is expected to be about $3.87 per ton of waste, and will include baling the waste. The city currently spends $0.061/ton-min for direct-haul to the landfill, but this cost will continue to increase as the city population grows and new development occurs. The city will spend $0.016/ton-min for hauling the baled waste from the transfer station to the existing landfill. The average travel time between the landfill and the transfer station or collection route is 72 min. At what direct-haul cost will it be more economical to build and operate the transfer station than to continue with direct-haul?

(A) $0.038/ton-min
(B) $0.054/ton-min
(C) $0.070/ton-min
(D) $0.11/ton-min

Hint: Find the breakeven point.

PROBLEM 100

A rural county has roll-off boxes for solid waste collection at 34 locations. The average location includes three boxes each with an 18 yd^3 capacity. The boxes are emptied twice monthly. Average driving time between any two locations is 27 min and the average travel time from any location to the landfill is 45 min.

The county is considering using smaller, 6 yd^3 dumpsters that can be emptied into a compaction truck. The compaction truck capacity is 12 yd^3 with a compaction factor of 3.0. For the same collection schedule, approximately how much driving time will be saved if compaction trucks and dumpsters are used instead of the roll-off boxes?

(A) 36 hr/mo
(B) 61 hr/mo
(C) 69 hr/mo
(D) 140 hr/mo

Hint: How many dumpsters are equivalent to a roll-off box?

Breadth Solutions

HYDRAULICS—CLOSED CONDUIT

SOLUTION 1

Because specific information regarding the type of pipe is not given, assume pipe diameters given are actual inside diameters.

A_1, A_2 cross-sectional area of upstream and downstream pipe, respectively in^2
d_1, d_2 inside diameter of upstream and downstream pipe, respectively in

$$A_1 = \frac{\pi d_1^2}{4} = \frac{\pi (2 \text{ in})^2}{4} = 3.14 \text{ in}^2$$

$$A_2 = \frac{\pi d_2^2}{4} = \frac{\pi (1.5 \text{ in})^2}{4} = 1.77 \text{ in}^2$$

v_1 water velocity in upstream pipe ft/sec
v_2 water velocity in downstream pipe ft/sec

$$v_2 = \frac{A_1 v_1}{A_2} = \frac{(3.14 \text{ in}^2)\left(4.2 \, \frac{\text{ft}}{\text{sec}}\right)}{1.77 \text{ in}^2} = 7.5 \text{ ft/sec}$$

The answer is (C).

Why Other Options Are Wrong

(A) This incorrect solution reverses the values for area in the velocity equation. Other definitions and equations are the same as used in the correct solution.

(B) This incorrect solution calculates circumference instead of area in the area equations. Other definitions and equations are the same as used in the correct solution.

(D) This incorrect solution makes a unit conversion error in the velocity equation and inverts the areas. Other definitions and equations are the same as used in the correct solution.

SOLUTION 2

The equivalent lengths for the fittings are summarized in the table. The equivalent lengths were taken from standard reference tables.

fitting	quantity	unit fitting equivalent length (ft)	total fitting equivalent length (ft)
90° ell	7	8.5	59.5
couple	16	0.45	7.2
union	4	0.45	1.8
tee	5	7.7	38.5
valve	2	1.5	3.0
			110

L length of pipe ft
L_e total equivalent length ft
L_f equivalent length of fittings ft

$$L_e = L + \sum L_f$$
$$= 250 \text{ ft} + 110 \text{ ft}$$
$$= 360 \text{ ft}$$

The answer is (D).

Why Other Options Are Wrong

(A) This incorrect solution ignores the pipe length in the equivalent length calculation. Other definitions are unchanged from the correct solution.

(B) This incorrect solution uses inches instead of feet as the units for fitting equivalent lengths. Other definitions are unchanged from the correct solution.

(C) This incorrect solution uses equivalent lengths for branched flow tees and long radius elbows. Other definitions are unchanged from the correct solution.

SOLUTION 3

Inside diameters for schedule-40 steel pipe are found in standard references.

pipe	inside diameter (in)	area (ft²)
A	2.469	0.0332
B	1.610	0.0141
C	1.049	0.00600

A pipe cross-sectional area ft²
Q water flow rate ft³/sec
v water velocity ft/sec

$$v = \frac{Q}{A}$$

$$v_A = \frac{0.76 \; \frac{ft^3}{sec}}{0.0332 \; ft^2} = 23 \; ft/sec$$

g gravitational acceleration 32.2 ft/sec²
p pressure lbf/ft²
γ water specific weight assume 62.4 lbf/ft³

Because no elevation data are given, assume that the elevation head is equal at all points for all three pipes. For such a case, the energy equation between pipes A and B is

$$\frac{p_A}{\gamma} + \frac{v_A^2}{2g} = \frac{p_B}{\gamma} + \frac{v_B^2}{2g}$$

$$\frac{\left(100 \; \frac{lbf}{in^2}\right)\left(144 \; \frac{in^2}{ft^2}\right)}{62.4 \; lbf/ft^3} + \frac{\left(23 \; \frac{ft}{sec}\right)^2}{(2)\left(32.2 \; \frac{ft}{sec^2}\right)}$$

$$= \frac{\left(90 \; \frac{lbf}{in^2}\right)\left(144 \; \frac{in^2}{ft^2}\right)}{62.4 \; \frac{lbf}{ft^3}} + \frac{v_B^2}{(2)\left(32.2 \; \frac{ft}{sec^2}\right)}$$

$$v_B = 45 \; ft/sec$$

The continuity equation is

$$A_A v_A = A_B v_B + A_C v_C$$

$$(0.0332 \; ft^2)\left(23 \; \frac{ft}{sec}\right) = (0.0141 \; ft^2)\left(45 \; \frac{ft}{sec}\right) + (0.00600 \; ft^2) v_C$$

$$v_C = 22 \; ft/sec$$

$$Q = vA$$

$$Q_C = \left(22 \; \frac{ft}{sec}\right)(0.00600 \; ft^2)$$

$$= 0.13 \; ft^3/sec$$

The answer is (A).

Why Other Options Are Wrong

(B) This incorrect solution miscalculates the energy equation. Other assumptions, definitions, and equations are the same as used in the correct solution.

(C) This incorrect solution ignores the continuity equation. The solution requires an assumption regarding the pressure in pipe C. Other assumptions, definitions, and equations are the same as used in the correct solution.

(D) This incorrect solution misuses the continuity equation and ignores the energy equation. Other assumptions, definitions, and equations are the same as used in the correct solution.

SOLUTION 4

For elevation head loss, the elevation difference does not contribute to head loss in the pipe. Therefore, head loss in the pipe occurs from pipe characteristics and fittings.

For equivalent length of threaded 2 in schedule-80 steel fittings, the following characteristics apply.

fitting	quantity	unit equivalent length (ft)	total equivalent length (ft)
couple	15	0.45	6.75
90° ell	8	8.5	68.0
45° ell	4	2.7	10.8
straight tee	6	7.7	46.2
globe valve	2	54	108
			239.75

For friction head loss in the pipe (including equivalent length of fittings),

D inside diameter for 2 in schedule-80 steel pipe
ε roughness coefficient for steel pipe

$$D = 1.939 \; in$$
$$\varepsilon = 0.0002 \; ft$$

$\frac{\varepsilon}{D}$ relative roughness –

$$\frac{\varepsilon}{D} = \frac{(0.0002 \; ft)\left(12 \; \frac{in}{ft}\right)}{1.939 \; in}$$
$$= 0.00124$$

Re Reynolds number –
v flow velocity ft/sec
ν kinematic viscosity ft²/sec

$\nu = 1.217 \times 10^{-5}$ ft/sec at 60°F

$$\text{Re} = \frac{D\text{v}}{\nu} = \frac{(1.939 \text{ in})\left(\frac{1 \text{ ft}}{12 \text{ in}}\right)\left(5 \frac{\text{ft}}{\text{sec}}\right)}{1.217 \times 10^{-5} \frac{\text{ft}^2}{\text{sec}}}$$

$$= 6.6 \times 10^4$$

f friction factor   –

$$f = \frac{0.25}{\left(\log\left(\frac{\frac{\varepsilon}{D}}{3.7} + \frac{5.74}{\text{Re}^{0.9}}\right)\right)^2}$$

$$= \frac{0.25}{\left(\log\left(\frac{0.00124}{3.7} + \frac{5.74}{(6.6 \times 10^4)^{0.9}}\right)\right)^2}$$

$$= 0.024$$

Note that the friction factor could also be determined using a Moody diagram with the same input values for Reynolds number and relative roughness, Re and ε/D.

g gravitational acceleration   ft/sec^2
h total head loss   ft
h_f head loss from friction   ft
L pipe length   ft
L_e equivalent pipe length of fittings   ft

$$g = 32.2 \text{ ft/sec}^2$$

$$h_f = \frac{f(L+L_e)\text{v}^2}{2Dg}$$

$$= \frac{(0.024)(500 \text{ ft} + 239.75 \text{ ft})\left(5 \frac{\text{ft}}{\text{sec}}\right)^2}{(2)(1.939 \text{ in})\left(\frac{1 \text{ ft}}{12 \text{ in}}\right)\left(32.2 \frac{\text{ft}}{\text{sec}^2}\right)}$$

$$= 43 \text{ ft}$$

The answer is (B).

Why Other Options Are Wrong

(A) This incorrect solution assumes that minor losses are negligible. Other assumptions, definitions, and equations are unchanged from the correct solution.

(C) This incorrect solution uses pressure drop tables for schedule-40 steel pipe. Other assumptions, definitions, and equations are unchanged from the correct solution.

(D) This incorrect solution assumes the equivalent length of the fittings represents the minor losses in feet of water. Other assumptions, definitions, and equations are unchanged from the correct solution.

SOLUTION 5

Assume that friction head loss is negligible.

g gravitational acceleration   32.2 ft/sec^2
h pressure head   ft
p pressure   psig or lbf/in^2
ρ water density at given temperature   lbm/ft^3

$$h = \frac{pg_c}{g\rho} = \frac{\left(100 \frac{\text{lbf}}{\text{in}^2}\right)\left(32.2 \frac{\text{ft-lbm}}{\text{lbf-sec}^2}\right)\left(144 \frac{\text{in}^2}{\text{ft}^2}\right)}{\left(32.2 \frac{\text{ft}}{\text{sec}^2}\right)\left(62.3 \frac{\text{lbm}}{\text{ft}^3}\right)}$$

$$= 231 \text{ ft}$$

Assume that the Torricelli equation applies.

C_v nozzle coefficient   –
v   jet velocity   ft/sec

$$\text{v} = C_\text{v}\sqrt{2gh} = (0.98)\sqrt{(2)\left(32.2 \frac{\text{ft}}{\text{sec}^2}\right)(231 \text{ ft})}$$

$$= 120 \text{ ft/sec}$$

A area of nozzle opening   ft^2
D nozzle opening diameter   ft

$$A = \pi\frac{D^2}{4} = \frac{\pi(0.5 \text{ in})^2\left(\frac{1 \text{ ft}^2}{144 \text{ in}^2}\right)}{4}$$

$$= 0.0014 \text{ ft}^2$$

Q flow rate   gal/min

$$Q = A\text{v}$$

$$= (0.0014 \text{ ft}^2)\left(120 \frac{\text{ft}}{\text{sec}}\right)\left(\frac{1 \text{ gal}}{0.134 \text{ ft}^3}\right)\left(60 \frac{\text{sec}}{\text{min}}\right)$$

$$= 75 \text{ gal/min}$$

The answer is (B).

Why Other Options Are Wrong

(A) This incorrect solution does not account for the difference between lbf and lbm when converting pressure to head. Other assumptions, definitions, and equations are the same as used in the correct solution.

(C) This incorrect solution fails to properly convert in^2 to ft^2 in the pressure head and nozzle area equations. Other assumptions, definitions, and equations are the same as used in the correct solution.

(D) This incorrect solution takes the square root of the head loss only, instead of the head loss-gravity constant term. Other assumptions, definitions, and equations are the same as used in the correct solution.

SOLUTION 6

- Q flow rate   gal/min
- ω operation (rotating) speed   rev/min

For the same pump operated at different speeds but with constant head and efficiency, the flow rates are proportional to the operating speed.

$$\frac{Q_1}{\omega_1} = \frac{Q_2}{\omega_2}$$

$$Q_2 = \frac{Q_1 \omega_2}{\omega_1} = \frac{\left(850\ \frac{\text{gal}}{\text{min}}\right)\left(2200\ \frac{\text{rev}}{\text{min}}\right)}{1750\ \frac{\text{rev}}{\text{min}}}$$

$$= 1069\ \text{gal/min} \quad (1100\ \text{gal/min})$$

The answer is (D).

Why Other Options Are Wrong

(A) This incorrect solution reduces the pumping rate by the efficiency and confuses the proportionality equation. Other definitions and equations are unchanged from the correct solution.

(B) This incorrect solution confuses the proportionality equation. Other definitions and equations are unchanged from the correct solution.

(C) This incorrect solution reduces the pumping rate by the efficiency. Other definitions and equations are unchanged from the correct solution.

SOLUTION 7

From the illustration, for a capacity of 75 gal/min, the net positive suction head is 14 ft.

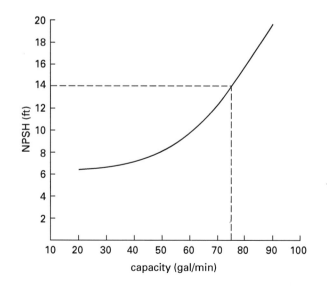

- h_f friction and minor head losses   ft
- NPSH   net positive suction head   ft
- p_o atmospheric pressure at 4573 ft above MSL   12.2 lbf/in^2
- p_v vapor pressure of water at 48°F   0.178 lbf/in^2
- z_s pump height above the water surface   ft
- γ specific weight of water at 48°F   62.4 lbf/ft^3

$$z_s = \left(\left(\frac{12.2\ \frac{\text{lbf}}{\text{in}^2}}{62.4\ \frac{\text{lbf}}{\text{ft}^3}}\right)\left(144\ \frac{\text{in}^2}{\text{ft}^2}\right)\right) - 14\ \text{ft}$$

$$- 6.7\ \text{ft} - \left(\left(\frac{0.178\ \frac{\text{lbf}}{\text{in}^2}}{62.4\ \frac{\text{lbf}}{\text{ft}^3}}\right)\left(144\ \frac{\text{in}^2}{\text{ft}^2}\right)\right)$$

$$= 7.0\ \text{ft}$$

- z_p maximum pump elevation   ft

$$z_p = 4573\ \text{ft} + 7.0\ \text{ft} = 4580\ \text{ft}$$

The answer is (B).

Why Other Options Are Wrong

(A) This incorrect solution subtracts the pump height from the water surface elevation instead of adding to it. The figure and other assumptions, definitions, and equations are unchanged from the correct solution.

(C) This incorrect solution uses the atmospheric pressure at sea level instead of at 4573 ft. The figure and other assumptions, definitions, and equations are unchanged from the correct solution.

(D) This incorrect solution adds instead of subtracts the net positive suction head. The figure and other assumptions, definitions, and equations are unchanged from the correct solution.

SOLUTION 8

- C Chezy coefficient   –
- D pipe diameter   m
- f friction factor   –
- g gravitational acceleration   9.81 m/s^2
- R hydraulic radius   m
- Re   Reynolds number   –
- S slope   m/m
- v flow velocity   m/s
- ε specific roughness 1.5×10^{-6} m typical for PVC pipe
- ε/D relative roughness   –
- ν kinematic viscosity 1.371×10^{-6} m^2/s at 10°C

An equation is needed to relate the friction factor to other known values. The required equation can be developed as follows.

$$C = \sqrt{\frac{8g}{f}}$$

$$v = C\sqrt{R}\sqrt{S} = \sqrt{\frac{8g}{f}}\sqrt{R}\sqrt{S}$$

$$\text{Re} = \frac{4Rv}{\nu} = \frac{4R\left(C\sqrt{R}\sqrt{S}\right)}{\nu}$$

$$= \frac{4R\sqrt{\frac{8g}{f}}\sqrt{R}\sqrt{S}}{\nu}$$

$$\frac{1}{\sqrt{f}} = -2\log\left(\frac{\frac{\varepsilon}{D}}{3.7} + \frac{2.51}{\text{Re}\sqrt{f}}\right)$$

$$= -2\log\left(\frac{\frac{\varepsilon}{D}}{3.7} + \frac{2.51\nu}{4R\sqrt{\frac{8g}{f}}\sqrt{R}\sqrt{S}\sqrt{f}}\right)$$

$$= -2\log\left(\frac{\frac{\varepsilon}{D}}{3.7} + \frac{0.22\nu}{\sqrt{g}R^{3/2}\sqrt{S}}\right)$$

$$\frac{\varepsilon}{D} = \frac{1.5 \times 10^{-6} \text{ m}}{(100 \text{ mm})\left(\frac{1 \text{ m}}{10^3 \text{ mm}}\right)}$$

$$= 0.000015$$

d flow depth m

The hydraulic radius can be determined from reference tables using the ratio of flow depth to pipe diameter.

$$\frac{d}{D} = \frac{43 \text{ mm}}{100 \text{ mm}} = 0.43$$

$$R = 0.023 \text{ m}$$

$$\frac{1}{\sqrt{f}} = -2\log\left(\frac{0.000015}{3.7} + \frac{(0.22)\left(1.371 \times 10^{-6} \frac{\text{m}^2}{\text{s}}\right)}{\sqrt{9.81 \frac{\text{m}}{\text{s}^2}}(0.023 \text{ m})^{3/2} \times \sqrt{0.03 \frac{\text{m}}{\text{m}}}}\right)$$

$$= 7.57$$

$$f = 0.0174 \quad (0.017)$$

The answer is (B).

Why Other Options Are Wrong

(A) This incorrect option includes a mathematical error in the development of the friction loss equation. Other assumptions, definitions, and equations are the same as used in the correct solution.

(C) This incorrect option fails to multiply by two after taking the log in the friction factor equation. Other assumptions, definitions, and equations are the same as used in the correct solution.

(D) This incorrect option uses the value for absolute viscosity with the kinematic viscosity units. Other assumptions, definitions, and equations are the same as used in the correct solution.

HYDRAULICS—OPEN CHANNEL

SOLUTION 9

A_o area of gate opening ft^2
C_d gate discharge coefficient –
g gravitational acceleration 32.2 ft/sec^2
h available head at the gate ft
Q discharge through the gate ft^3/sec

Because a partially open gate has the characteristics of a submerged orifice, the following equation applies.

$$A_o = \frac{Q}{C_d\sqrt{2gh}} = \frac{80 \frac{\text{ft}^3}{\text{sec}}}{(0.72)\sqrt{(2)\left(32.2 \frac{\text{ft}}{\text{sec}^2}\right)(4.3 \text{ ft})}}$$

$$= 6.7 \text{ ft}^2$$

The answer is (C).

Why Other Options Are Wrong

(A) This incorrect solution fails to take the square root of the term in the denominator. Definitions are unchanged from the correct solution.

(B) This incorrect solution applies the square root to the head only, instead of the head and gravitational acceleration term. Definitions are unchanged from the correct solution.

(D) This incorrect solution uses the equation for flow over a weir instead of through a submerged orifice. This requires an erroneous assumption that the gate area is equal to the product of the upstream head and the gate width. Definitions are unchanged from the correct solution.

SOLUTION 10

For a unit length of smooth earthen channel, it is reasonable to assume that flow is uniform and that all head loss is due to friction. With these assumptions, the head loss is the product of the channel length and slope, with the slope determined using the Manning equation.

b	channel base width	m
d	water depth	m
R	hydraulic radius	m
θ	angle from the horizontal to the side wall	degree

For 1-to-1 side slopes, the angle from the horizontal to the side wall is 45°.

$$R = \frac{bd\sin\theta + d^2\cos\theta}{b\sin\theta + 2d}$$
$$= \frac{(2\text{ m})(0.5\text{ m})(\sin 45°) + (0.5\text{ m})^2(\cos 45°)}{(2\text{ m})(\sin 45°) + (2)(0.5\text{ m})}$$
$$= 0.366\text{ m}$$

h_f	head loss from friction	m/unit length of channel
L	channel length	m
n	Manning roughness coefficient	–
v	water-flow velocity	m/s

Use 1 m for the unit length of the channel. The Manning roughness coefficient for a smooth earthen lining is 0.018.

$$h_f = \frac{Ln^2 v^2}{R^{4/3}} = \frac{(1\text{ m})(0.018)^2 \left(3.2\,\frac{\text{m}}{\text{s}}\right)^2}{(0.366\text{ m})^{4/3}}$$
$$= 0.013\text{ m/m length of channel}$$

The answer is (C).

Why Other Options Are Wrong

(A) This incorrect solution calculates the wetted perimeter instead of the hydraulic radius for the channel. Other assumptions, definitions, and equations are unchanged from the correct solution.

(B) This incorrect solution fails to square the velocity term in the head loss equation. Other assumptions, definitions, and equations are unchanged from the correct solution.

(D) This incorrect solution uses the Manning roughness coefficient for natural channels instead of smooth earth. Other assumptions, definitions, and equations are unchanged from the correct solution.

SOLUTION 11

As the water flows over the initial break from horizontal to a relatively steep slope and the flow transitions from subcritical to supercritical, the water depth will likely be at critical depth at the break in slope. Water will flow at less than critical depth (supercritical velocity) through both steeply sloped sections. At the bottom of the second steep section where the slope changes from relatively steep to relatively flat, a hydraulic jump will form as flow transitions from supercritical back to subcritical. The hydraulic jump will form close to the slope transition. The hump in the channel bottom could influence flow in a variety of ways, but its relative distance from the steep to shallow slope change and occurrence after the hydraulic jump where the water is deeper will limit this influence.

The answer is (B).

Why Other Options Are Wrong

(A) The flow profile generally follows what would be expected until it reaches the transition from the steep to nearly horizontal section. The hydraulic jump would likely form near the transition, some distance upstream of the hump instead of, as the illustration shows, at the hump.

(C) The flow profile would likely show critical flow beginning at the top of the first steep section and a hydraulic jump would likely occur at the bottom of the second steep section. This illustration does not show either of these features.

(D) Because of the slope change from relatively steep to relatively flat, it is unlikely that the flow would be subcritical in the upstream horizontal section and then continue as subcritical flow after the slope transition as shown in the illustration.

SOLUTION 12

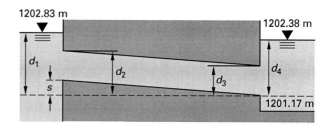

d_3	culvert diameter	m
L	culvert length	m
S	slope	m/m

Other terms are defined in the illustration. All have units of m.

$$d_1 = 1202.83 \text{ m} - 1201.17 \text{ m} = 1.66 \text{ m}$$

$$s = SL = \left(\frac{2 \text{ m}}{100 \text{ m}}\right)(27 \text{ m})$$
$$= 0.54 \text{ m}$$

$$d_2 = d_3 + s = 1 \text{ m} + 0.54 \text{ m}$$
$$= 1.54 \text{ m}$$

Because d_1 is greater than d_2, the culvert inlet is submerged.

$$d_4 = 1202.38 \text{ m} - 1201.17 \text{ m} = 1.21 \text{ m}$$
$$d_3 = 1 \text{ m}$$

Because d_4 is greater than d_3, the culvert outlet is submerged.

Submerged inlet and outlet classify the culvert flow as type-4, submerged outlet.

The answer is (B).

Why Other Options Are Wrong

(A) This incorrect solution confuses elevations used when calculating the inlet and outlet water depths, resulting in an exposed (not submerged) inlet and free flow outlet, which classify the culvert flow as type-3, tranquil flow. Definitions are the same as used in the correct solution.

(C) This incorrect solution checks the inlet conditions only. Submerged inlet classifies the culvert flow as type-5, rapid flow, at inlet. Definitions are the same as used in the correct solution.

(D) This incorrect solution confuses the elevation values when calculating the outlet water depth, resulting in a submerged inlet and free-flow outlet, which classify the culvert flow as type-6, full flow, free outlet. Definitions are the same as used in the correct solution.

SOLUTION 13

d_1 upstream water depth   m
Q water flow   m³/s
v_1 upstream water velocity   m/s
w channel width   m

$$v_1 = \frac{Q}{d_1 w} = \frac{2.93 \ \frac{\text{m}^3}{\text{s}}}{(0.15 \text{ m})(2.5 \text{ m})} = 7.8 \text{ m/s}$$

d_2 downstream water depth   m
g gravitational acceleration   9.81 m/s²
v_2 downstream water velocity   m/s

$$d_2 = -(0.5)d_1 + \sqrt{\frac{2v_1^2 d_1}{g} + 0.25 d_1^2}$$

$$= -(0.5)(0.15 \text{ m}) + \sqrt{\left(\frac{(2)\left(7.8 \ \frac{\text{m}}{\text{s}}\right)^2 (0.15 \text{ m})}{9.81 \ \frac{\text{m}}{\text{s}^2}}\right) + (0.25)(0.15 \text{ m})^2}$$

$$= 1.29 \text{ m}$$

$$v_2 = \frac{Q}{d_2 w} = \frac{2.93 \ \frac{\text{m}^3}{\text{s}}}{(1.29 \text{ m})(2.5 \text{ m})}$$
$$= 0.91 \text{ m/s}$$

ΔH head dissipated   m

$$\Delta H = d_1 + \frac{v_1^2}{2g} - d_2 - \frac{v_2^2}{2g}$$
$$= 0.15 \text{ m}$$
$$+ \frac{\left(7.8 \ \frac{\text{m}}{\text{s}}\right)^2}{(2)\left(9.81 \ \frac{\text{m}}{\text{s}^2}\right)} - 1.29 \text{ m} - \frac{\left(0.91 \ \frac{\text{m}}{\text{s}}\right)^2}{(2)\left(9.81 \ \frac{\text{m}}{\text{s}^2}\right)}$$
$$= 1.92 \text{ m} \quad (1.9 \text{ m})$$

The answer is (C).

Why Other Options Are Wrong

(A) This incorrect solution fails to take the square root in the downstream depth calculation. Other definitions and equations are unchanged from the correct solution.

(B) This incorrect solution adds instead of subtracts the upstream depth term in the downstream depth calculation. Other definitions and equations are unchanged from the correct solution.

(D) This incorrect solution reverses the values for upstream and downstream depths in the energy dissipation equation. Other definitions and equations are unchanged from the correct solution.

SOLUTION 14

b	channel bottom width	m
d_c	critical water depth	m
w	channel width at water surface	m
θ	side slope angle measured from the horizontal	degree

For a trapezoidal channel with 1-to-1 side slopes, the side slope angle is 45°.

$$w = b + 2d_c \tan\theta = 8\text{ m} + 2d_c \tan 45°$$
$$= 8\text{ m} + 2d_c$$

A	channel cross-sectional area	m^2
Fr	Froude number	–
g	gravitational acceleration	9.81 m/s^2
Q	flow rate	m^3/s

At critical flow for a non-rectangular channel,

$$\text{Fr} = 1 = \sqrt{\frac{Q^2 w}{gA^3}}$$

Solve for A in terms of d_c using Froude equation.

$$A = \left(\frac{Q^2 w}{g}\right)^{1/3} = \left(\frac{\left(22\,\frac{\text{m}^3}{\text{s}}\right)^2 w}{9.81\,\frac{\text{m}}{\text{s}^2}}\right)^{1/3}$$
$$= 3.7 w^{1/3}$$
$$= (3.7)(8\text{ m} + 2d_c)^{1/3}$$

Solve for A in terms of d_c using channel dimensions.

$$A = bd_c + \frac{d_c^2}{\tan\theta} = (8\text{ m})d_c + \frac{d_c^2}{\tan 45°}$$
$$= (8\text{ m})d_c + d_c^2$$

Solve for d_c using results from above calculations for A.

$$(8\text{ m})d_c + d_c^2 = (3.7)(8\text{ m} + 2d_c)^{1/3}$$
$$d_c = 0.89\text{ m}$$
$$A = (8\text{ m})(0.89\text{ m}) + (0.89)^2$$
$$= 7.9\text{ m}^2$$

R hydraulic radius m

$$R = \frac{bd_c \sin\theta + d_c^2 \cos\theta}{b\sin\theta + 2d_c}$$
$$= \frac{(8\text{ m})(0.89\text{ m})(\sin 45°) + (0.89\text{ m})^2(\cos 45°)}{(8\text{ m})(\sin 45°) + (2)(0.89\text{ m})}$$
$$= 0.75\text{ m}$$

n Manning roughness coefficient –
S channel slope m/m

Use 0.013 for the Manning roughness coefficient for concrete lining.

$$\frac{Q}{A} = \left(\frac{1}{n}\right) R^{2/3} \sqrt{S}$$

$$S = \left(\frac{Qn}{AR^{2/3}}\right)^2 = \left(\frac{\left(22\,\frac{\text{m}^3}{\text{s}}\right)(0.013)}{(7.9\text{ m}^2)(0.75\text{ m})^{2/3}}\right)^2$$
$$= 0.0019\text{ m/m}$$

The answer is (C).

Why Other Options Are Wrong

(A) This incorrect solution uses the equation for the wetted perimeter as the equation for the hydraulic radius. Other assumptions, definitions, and equations are unchanged from the correct solution.

(B) This incorrect solution makes an error when calculating the area using the Froude number equation. Other assumptions, definitions, and equations are unchanged from the correct solution.

(D) This incorrect solution miscalculates the channel width at the water surface. Other assumptions, definitions, and equations are unchanged from the correct solution.

SOLUTION 15

This problem can be solved using either conjugate depths or the specific force equation. This solution uses the specific force equation.

A	cross-sectional flow area	m^2
b	channel base width	m
d	water depth	m
θ	side slope angle measured from the horizontal	–
1, 2 subscripts	upstream and downstream, respectively	

For a trapezoidal channel,

$$A = bd + \frac{d^2}{\tan\theta}$$

For 1-to-1 side slopes, the side slope angle is 45°.

$$A_1 = (4.2\text{ m})(0.74\text{ m}) + \frac{(0.74\text{ m})^2}{\tan 45°} = 3.66\text{ m}^2$$
$$A_2 = (4.2\text{ m})d_2 + d_2^2$$

$\bar{d}$ distance from centroid to water surface m

$$A\bar{d} = \frac{d^2\left(3b + \frac{2d}{\tan\theta}\right)}{6}$$

$$A_1\bar{d}_1 = \frac{(0.74\text{ m})^2\left((3)(4.2\text{ m}) + \frac{(2)(0.74\text{ m})}{\tan 45°}\right)}{6}$$

$$= 1.29\text{ m}^3$$

$$A_2\bar{d}_2 = \frac{d_2^2\left((3)(4.2\text{ m}) + \frac{(2)d_2}{\tan 45°}\right)}{6}$$

$$= d_2^2(2.1\text{ m} + 0.33d_2)$$

g gravitational acceleration   9.81 m/s²
Q flow rate   m³/s

$$A_1\bar{d}_1 + \frac{Q^2}{gA_1} = A_2\bar{d}_2 + \frac{Q^2}{gA_2}$$

$$1.29\text{ m}^3 + \frac{\left(38\,\frac{\text{m}^3}{\text{s}}\right)^2}{\left(9.81\,\frac{\text{m}}{\text{s}^2}\right)(3.66\text{ m}^2)}$$

$$= d_2^2(2.1\text{ m} + 0.33d_2)$$
$$+ \frac{\left(38\,\frac{\text{m}^3}{\text{s}}\right)^2}{\left(9.81\,\frac{\text{m}}{\text{s}^2}\right)((4.2\text{ m})d_2 + d_2^2)}$$

$$41.5 = 2.1d_2^2 + 0.33d_2^3 + \frac{147}{4.2d_2 + d_2^2}$$

$$d_2 = 3.3\text{ m}$$

The answer is (D).

Why Other Options Are Wrong

(A) This incorrect solution squares both the flow rate and area terms in the specific force equation calculation. Other assumptions, definitions, and equations are unchanged from the correct solution.

(B) This incorrect solution uses the geometry and equations for a rectangular channel and assumes that the centroid depth and water depth are equal. Other assumptions, definitions, and equations are unchanged from the correct solution.

(C) This incorrect solution assumes that the centroid depth and the water depth are equal. Other assumptions, definitions, and equations are unchanged from the correct solution.

HYDROLOGY

SOLUTION 16

section	velocity at 0.2 depth (m/s)	velocity at 0.8 depth (m/s)	average velocity (m/s)
AB	–	–	–
BC	0.41	0.32	0.365
CD	0.44	0.32	0.38
DE	0.48	0.34	0.41
EF	0.48	0.33	0.405
FG	0.49	0.36	0.425
GH	0.49	0.35	0.42
HI	0.51	0.37	0.44
IJ	0.50	0.36	0.43
JK	0.52	0.37	0.445
KL	0.50	0.38	0.44
LM	0.49	0.35	0.42
MN	0.50	0.36	0.43
NO	0.47	0.34	0.405
OP	0.43	0.31	0.37
PQ	0.41	0.32	0.365
QR	0.39	0.30	0.345
RS	–	–	–

section	average depth (m)	section width (m)	flow area (m²)	flow rate (m³/s)
AB	0.7	1.2	0.84	–
BC	1.9	1.1	2.09	0.76
CD	2.3	1.3	2.99	1.14
DE	2.7	1.3	3.51	1.44
EF	2.9	1.2	3.48	1.41
FG	3.0	1.1	3.30	1.40
GH	3.1	1.1	3.41	1.43
HI	2.9	1.2	3.48	1.53
IJ	2.9	1.2	3.48	1.50
JK	2.8	1.4	3.92	1.74
KL	3.1	1.2	3.72	1.64
LM	2.8	1.1	3.08	1.29
MN	2.7	1.1	2.97	1.28
NO	2.5	1.3	3.25	1.32
OP	2.0	1.2	2.40	0.89
PQ	1.8	1.2	2.16	0.79
QR	1.6	1.2	1.92	0.66
RS	0.5	1.2	0.60	–
				20.22

The tabulated calculations are from the following equations.

v section average flow velocity   m/s
v_2 section velocity at 0.2 depth   m/s
v_8 section velocity at 0.8 depth   m/s

$$v = \frac{v_2 + v_8}{2}$$

A	section flow area	m²
d	section average flow depth	m
w	section flow width	m

$$A = dw$$

q	section flow rate	m³/s

$$q = Av$$

Q	total flow rate	m³/s

$$Q = \sum q$$
$$= 20.22 \text{ m}^3/\text{s} \quad (20 \text{ m}^3/\text{s})$$

The answer is (B).

Why Other Options Are Wrong

(A) This incorrect solution calculates the average flow in each section instead of the total flow. The table and other definitions and equations are unchanged from the correct solution.

(C) This incorrect solution uses the sum of the velocities at each depth fraction instead of calculating the average velocity. Other definitions and equations are unchanged from the correct solution.

(D) This incorrect solution uses the total area and the total average velocity to calculate flow rate. Other definitions and equations are unchanged from the correct solution.

SOLUTION 17

d	rainfall depth	in
t	storm duration	hr

From the illustration, for mean annual precipitation of 27 in and storm duration of 2.5 hr, the rainfall depth is about 1.5 in.

i	rainfall intensity	in/hr

$$i = \frac{d}{t} = \frac{1.5 \text{ in}}{2.5 \text{ hr}}$$
$$= 0.60 \text{ in/hr}$$

The answer is (B).

Why Other Options Are Wrong

(A) This incorrect choice misreads the illustration by selecting a value for rainfall depth between the 20 in and 25 in mean annual precipitation. Definitions and equations are unchanged from the correct solution.

(C) This incorrect choice assumes the rainfall intensity is equal to the rainfall depth.

(D) This incorrect choice takes the ratio of the mean annual precipitation and the storm duration as the rainfall intensity. Definitions are unchanged from the correct solution.

SOLUTION 18

Assume runoff coefficients are the average values in the range of typical values listed in standard references.

land use	area (%)	runoff coefficient	land area (ac)
apartments	30	0.60	39.3
landscaped open space (park)	25	0.17	32.75
light industrial	45	0.65	58.95
			131

A	land area for each use (subscript j designates each land use)	ac
C	runoff coefficient for each land use (subscript ave designates average)	–

$$C_{\text{ave}} = \frac{\sum C_j A_j}{\sum A_j}$$
$$= \frac{\begin{pmatrix}(0.60)(39.3 \text{ ac}) + (0.17)(32.75 \text{ ac}) \\ + (0.65)(58.95 \text{ ac})\end{pmatrix}}{131 \text{ ac}}$$
$$= 0.515$$

i	storm intensity	in/hr

From the illustration, the storm intensity is about 6 in/hr.

Q	runoff	ac-ft/hr

$$Q = C_{\text{ave}} i A = (0.515)\left(6 \; \frac{\text{in}}{\text{hr}}\right)(131 \text{ ac})\left(\frac{1 \text{ ft}}{12 \text{ in}}\right)$$
$$= 34 \text{ ac-ft/hr}$$

The answer is (B).

Why Other Options Are Wrong

(A) This incorrect solution calculates the arithmetic average instead of the weighted average of the runoff coefficients. Assumptions, definitions, and equations are unchanged from the correct solution.

(C) This incorrect solution calculates the runoff for the total area using the summed runoff coefficients. Other assumptions, definitions, and equations are unchanged from the correct solution.

(D) This incorrect solution misreads the units for intensity from the figure. Assumptions, definitions, and equations are unchanged from the correct solution.

SOLUTION 19

I floods creating economic impact   %
n_e number of events
n_f number of floods

$$I = \left(\frac{n_f}{n_e}\right) \times 100\% = \left(\frac{18}{112}\right) \times 100\% = 16\%$$

Entering the figure at 16% on the y-axis and extending horizontally to intersect the curve, the peak flow is 1600 m³/s.

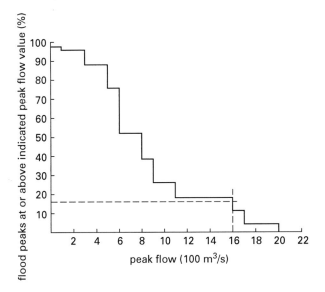

The answer is (C).

Why Other Options Are Wrong

(A) This incorrect solution determines the peak flow for the percentage of floods that did not result in significant economic impact. Other definitions are unchanged from the correct solution.

(B) This incorrect solution reads the number of events instead of the percentage of these events resulting in economic impact.

(D) This incorrect solution reads the x-axis scale as 1000 m³/s instead of 100 m³/s. The figure and other definitions and equations are unchanged from the correct solution.

SOLUTION 20

Because the normal annual precipitation between station C and the other three stations varies by more than 10%, the normal-ratio method should be used for estimating the missing record.

N normal annual precipitation   cm
P annual precipitation for year of interest   cm

Subscripts A, B, C, and D refer to the precipitation stations.

$$P_C = \left(\frac{N_C}{3}\right)\left(\frac{P_A}{N_A} + \frac{P_B}{N_B} + \frac{P_D}{N_D}\right)$$
$$= \left(\frac{42 \text{ cm}}{3}\right)\left(\frac{34 \text{ cm}}{39 \text{ cm}} + \frac{28 \text{ cm}}{31 \text{ cm}} + \frac{32 \text{ cm}}{37 \text{ cm}}\right)$$
$$= 37 \text{ cm}$$

The answer is (B).

Why Other Options Are Wrong

(A) This incorrect solution takes the arithmetic average of the other three stations for 1991 as the missing record for station C in 1991. Definitions are unchanged from the correct solution.

(C) This incorrect solution misreads one of the values entered in the normal-ratio equation. Definitions and the equation are unchanged from the correct solution.

(D) This incorrect solution reverses the values for the normal annual precipitation and the annual precipitation in the year of interest. Other definitions and the equation are unchanged from the correct solution.

SOLUTION 21

The rainfall volume in surface detention is defined by the area to the left of the curve and under a horizontal line extended from the y-axis to the peak of the curve as shown in the following illustration.

Using the illustration, calculate the volume of each segment of grid area.

$$\left(25 \, \frac{\text{m}^3}{\text{s}}\right)(1 \text{ h})\left(3600 \, \frac{\text{s}}{\text{h}}\right) = 90\,000 \text{ m}^3$$

time interval (h)	segments
0–1	15.5
1–2	11.5
2–3	3.0
3–4	0.5
4–5	0.0
	30.5

I surface detention   m³

$$I = (30.5)(90\,000 \text{ m}^3) = 2.7 \times 10^6 \text{ m}^3$$

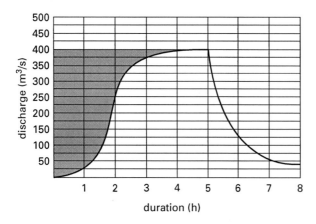

The answer is (B).

Why Other Options Are Wrong

(A) This incorrect solution calculates the volume as the area under the recession portion of the curve beginning at 5 h when rainfall ends. Definitions are unchanged from the correct solution.

(C) This incorrect solution calculates the volume as the area under the curve until the end of rainfall at 5 h. Definitions are unchanged from the correct solution.

(D) This incorrect solution calculates the volume based on the total area under the curve instead of to the left of it. Definitions are unchanged from the correct solution.

SOLUTION 22

Typical runoff coefficients are selected from standard reference tables. Where a range of typical values are listed in reference tables, an average of the range of values is used.

land use	area (ac)	typical runoff coefficient	area weighted runoff coefficient
shingle roof	0.8	0.85	0.68
concrete surface	1.1	0.88	0.97
asphalt surface	2.6	0.83	2.2
poorly drained lawn	10	0.15	1.5
	14.5		5.4

A watershed total area   ac
C_m watershed average weighted runoff coefficient   –
C_w area weighted runoff coefficient   –

$$C_m = \frac{\sum C_w}{A} = \frac{5.4}{14.5} = 0.37$$

A commonly used equation for runoff from urban areas is the FAA formula.

L longest runoff flow path   ft
S average ground surface slope   %
t_c time of concentration   min

$$t_c = \frac{(1.8)(1.1 - C_m)\sqrt{L}}{S^{1/3}}$$

$$= \frac{(1.8)(1.1 - 0.37)\sqrt{(337 \text{ yd})\left(3 \frac{\text{ft}}{\text{yd}}\right)}}{(0.013 \times 100\%)^{1/3}}$$

$$= 38 \text{ min}$$

The rainfall intensity can be estimated from the Steel formula with values for coefficients K and b taken for generalized rainfall regions and storm frequency. The typical coefficient values for the western United States and a 10-year storm are

b Steel formula constant   min
K Steel formula constant   in-min/hr

$$K = 60 \text{ in-min/hr}$$
$$b = 13 \text{ min}$$

i rainfall intensity   in/hr

$$i = \frac{K}{t_c + b} = \frac{60 \frac{\text{in-min}}{\text{hr}}}{38 \text{ min} + 13 \text{ min}} = 1.18 \text{ in/hr}$$

Q runoff flow rate   ft³/sec

$$Q = CiA = (0.37)\left(1.18 \frac{\text{in}}{\text{hr}}\right)(14.5 \text{ ac})\left(1 \frac{\text{ft}^3\text{-hr}}{\text{ac-in-sec}}\right)$$
$$= 6.3 \text{ ft}^3/\text{sec}$$

The answer is (B).

Why Other Options Are Wrong

(A) This incorrect solution uses the slope with units of foot per foot instead of as a percent. Other assumptions, definitions, and equations are unchanged from the correct solution.

(C) This incorrect solution reads the flow distance figure in feet instead of yards. Other assumptions, definitions, and equations are unchanged from the correct solution.

(D) This incorrect solution selects values for the Steel formula coefficients of 100 for K and 10 for b, assuming these to be reasonable approximate values. Other assumptions, definitions, and equations are unchanged from the correct solution.

SOLUTION 23

The Gumbel distribution can be used with the given information.

K_{50} Gumbel distribution coefficient for 50 yr record and 50-year flood   –

K_{83} Gumbel distribution coefficient for 83 yr record and 50-year flood   –

K_{100} Gumbel distribution coefficient for 100 yr record and 50-year flood   –

n years of record   –

From reference tables, the Gumbel distribution coefficient is available for record periods of 50 yr and 100 yr. Interpolation is required to find the Gumbel distribution coefficient for 83 yr.

$$K_{83} = K_{50} + \frac{(K_{100} - K_{50})(n - 50)}{100 - 50}$$
$$= 2.89 + \frac{(2.77 - 2.89)(83 - 50)}{100 - 50}$$
$$= 2.81$$

σ standard deviation   ft³/sec
Q_m average flow rate over period of record   ft³/sec
Q_p flow rate for the period of interest   ft³/sec

$$Q_p = Q_m + K_{83}\sigma = 1947\,\frac{\text{ft}^3}{\text{sec}} + (2.81)\left(613\,\frac{\text{ft}^3}{\text{sec}}\right)$$
$$= 3670\,\text{ft}^3/\text{sec} \quad (3700\,\text{ft}^3/\text{sec})$$

The answer is (B).

Why Other Options Are Wrong

(A) This incorrect solution takes the simple ratio of the average flow for the period of record to a 50 yr period. Other definitions are unchanged from the correct solution.

(C) This incorrect solution reads the value for the Gumbel coefficient for the return period instead of the record of length from the reference table. Other assumptions, definitions, and equations are unchanged from the correct solution.

(D) This incorrect solution multiplies the average flow instead of the standard deviation by the Gumbel coefficient. Other assumptions, definitions, and equations are unchanged from the correct solution.

WASTEWATER TREATMENT

SOLUTION 24

Express infiltration to each section as $m^3/d\cdot mm\cdot km$.

section	total infiltration to section (m^3/d)	pipe diameter (mm)	pipe length (km)	total unit infiltration ($m^3/d\cdot mm\cdot km$)	infiltration to section ($m^3/d\cdot mm\cdot km$)
1	2315	100	13.4	1.73	
		200	6.8	1.70	
		300	6.2	1.24	4.67
2	958	100	9.2	1.04	
		200	7.1	0.67	
		300	4.4	0.73	2.44
3	3996	100	24.9	1.60	
		200	12.1	1.65	
		300	11.9	1.12	4.37
4	1867	100	21.3	0.88	
		200	11.0	0.85	
		300	4.7	1.32	3.05

$$\text{total unit infiltration} = \frac{\text{total infiltration to section}}{(\text{pipe diameter})(\text{pipe length})}$$

The section with the highest infiltration rate when pipe diameter and pipe length are included is section 1. Section 1 should receive first priority for rehabilitation.

The answer is (A).

Why Other Options Are Wrong

(B) This choice is incorrect because section 2 has the lowest infiltration rate either based on units of m^3/d or units of $m^3/d\cdot mm\cdot km$. Although this may be the least expensive to rehabilitate because it has the shortest total pipe length, the contribution to infiltration from this section does not justify its rehabilitation ahead of other sections.

(C) This choice is incorrect because it selects, by observation of the infiltration data in units of m^3/d, the section with the highest overall infiltration rate, section 3.

(D) This choice is incorrect because section 4 does not present any characteristics that would place it ahead of other sections. It would be appropriate to rehabilitate section 4 before section 2, but not before section 1. Some economic justification based on total pipe length may exist for rehabilitating section 4 before section 3, but not before section 2.

SOLUTION 25

time period (hr)	period average flow (10^6 gal/day)	period flow volume (10^6 gal)	cumulative volume (10^6 gal)
0000–0400	1.39	0.232	0.232
0400–0800	3.21	0.535	0.767
0800–1200	4.05	0.675	1.442
1200–1600	2.63	0.438	1.880
1600–2000	3.91	0.652	2.532
2000–2400	1.98	0.330	2.862
			2.862

The time period represents a 4 hr interval.

Q	period average flow rate	10^6 gal/day
t	time period	hr
V	flow volume	10^6 gal
V_c	cumulative volume	10^6 gal

$$V = Qt = \frac{Q(4\text{ hr})}{24\,\frac{\text{hr}}{\text{day}}}$$

$$V_{c,i} = V_i + V_{c,i-1}$$

The daily average flow is

$$Q_a = \frac{2.862 \times 10^6 \text{ gal}}{6 \text{ periods}} = \frac{0.477 \times 10^6 \text{ gal}}{\text{period}}$$

The line plotted from the ordinate to the data point corresponding to 24 h represents the average flow. Lines representing the maximum deviation above and below the average flow are plotted parallel to the average flow line.

The volume represented by the deviation between the upper and lower parallel lines is the volume required for flow equalization as shown in the illustration.

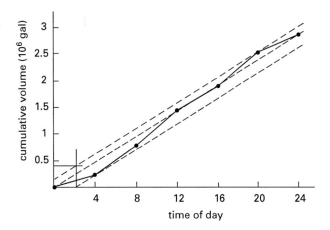

The tank volume is 5.0×10^5 gal.

The answer is (A).

Why Other Options Are Wrong

(B) This solution is incorrect because the required tank volume is calculated as the simple average of the cumulative volume. Other definitions are unchanged from the correct solution.

(C) This solution is incorrect because the required tank volume is determined using a plot of the cumulative average flow rate instead of the cumulative flow volume. To get volume from the illustration, the cumulative flow units are incorrectly taken as gallons. Other definitions are unchanged from the correct solution.

(D) This solution is incorrect because the required tank volume is determined from the simple average of the sum of the average flows. Notice that this is the total flow volume. Other definitions are unchanged from the correct solution.

SOLUTION 26

The pretreatment system's capacity should be able to meet the demand 90% of the time.

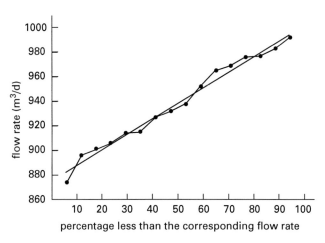

From the illustration, the desired flow rate = 990 m³/d.

The answer is (C).

Why Other Options Are Wrong

(A) This incorrect choice uses 90% of the average value of the average weekly flows.

(B) This incorrect choice uses the flow rate from the illustration at 10%.

(D) This incorrect choice divides the average of the average weekly flows by 90%.

WATER TREATMENT

SOLUTION 27

For gross production per filter,

n	number of filters	–
Q	total flow rate to filters	m³/d
q_n	flow rate per filter for all filters operating in filtration mode	m³/d

$$q_n = \frac{Q}{n} = \frac{28\,500 \, \frac{\text{m}^3}{\text{d}}}{5 \text{ filters}} = 5700 \text{ m}^3/\text{d}$$

For water lost to backwash,

A	filter bed area	m²
v_s	filtration rate	m³/m²·d

$$A = \frac{Q}{nv_s} = \frac{28\,500 \, \frac{\text{m}^3}{\text{d}}}{(5 \text{ filters})\left(225 \, \frac{\text{m}^3}{\text{m}^2\cdot\text{d}}\right)}$$

$$= 25 \text{ m}^2/\text{filter}$$

Q_b	backwash wastewater flow rate	m³/filter·d
v_b	backwash rate	m³/m²·d
m	number of daily backwash events/filter	1/d
t_b	time for backwash/filter	h

$$Q_b = v_b t_b A m$$

$$= \left(36 \, \frac{\text{m}^3}{\text{m}^2\cdot\text{h}}\right)(25 \text{ min})\left(25 \, \frac{\text{m}^2}{\text{filter}}\right)$$

$$\times \left(\frac{2}{\text{d}}\right)\left(\frac{\text{h}}{60 \text{ min}}\right)$$

$$= 750 \text{ m}^3/\text{filter}\cdot\text{d}$$

For water lost to conditioning,

t_c	time for conditioning/filter	h

$$Q_c = \frac{Qmt_c}{n} = \frac{\left(28\,500 \, \frac{\text{m}^3}{\text{d}}\right)\left(\frac{2}{\text{d}}\right)(8 \text{ min})}{(5 \text{ filters})\left(\frac{1440 \text{ min}}{\text{d}}\right)}$$

$$= 63 \text{ m}^3/\text{filter}\cdot\text{d}$$

Q_{net} net production per filter   m³/d

$$Q_{\text{net}} = q_n - Q_b - Q_c$$

$$= 5700 \, \frac{\text{m}^3}{\text{d}} - 750 \, \frac{\text{m}^3}{\text{d}} - 63 \, \frac{\text{m}^3}{\text{d}}$$

$$= 4887 \text{ m}^3/\text{d} \quad (4900 \text{ m}^3/\text{d})$$

The answer is (A).

Why Other Options Are Wrong

(B) This incorrect solution overcompensates for the increased hydraulic loading during backwashing and conditioning. Other definitions and equations are unchanged from the correct solution.

(C) This incorrect solution does not consider the increased filtration rate to the remaining filters when one filter is being backwashed and accounts for only one backwash event per filter per day. Other definitions and equations are unchanged from the correct solution.

(D) This incorrect solution fails to account for two backwash events per day. Other definitions and equations are unchanged from the correct solution.

SOLUTION 28

Because the Darcy friction factor and flow rate in units of ft³/sec were given, use the applicable Darcy equation in the following form.

f	Darcy friction factor	–
k	loss coefficient	–
L	pipe length	ft

$$k = \frac{0.025 fL}{D^5}$$

pipe	length (ft)	diameter (ft)	loss coefficient (unitless)
AB	3400	0.50	63
BC	2600	0.33	382
CA	1900	0.50	35

n	exponent for Darcy equation	–
Q	flow rate	ft³/sec
Δ	incremental flow rate change	ft³/sec

For the Darcy equation, use an exponent value of 2. Assume initial flows of 1.25 ft³/sec in pipe AB and 0.75 ft³/sec in pipes BC and CA. Assume positive flow is clockwise.

$$\Delta = \frac{-\sum kQ^n}{n\sum |kQ^{n-1}|}$$

Trial 1,

$$\Delta_1 = \frac{-\left(\begin{array}{c}(63)\left(1.25\,\dfrac{\text{ft}^3}{\text{sec}}\right)^2 - (382)\left(0.75\,\dfrac{\text{ft}^3}{\text{sec}}\right)^2 \\ -(35)\left(0.75\,\dfrac{\text{ft}^3}{\text{sec}}\right)^2\end{array}\right)}{(2)\left(\begin{array}{c}\left|(63)\left(1.25\,\dfrac{\text{ft}^3}{\text{sec}}\right)\right| + \left|(382)\left(0.75\,\dfrac{\text{ft}^3}{\text{sec}}\right)\right| \\ + \left|(35)\left(0.75\,\dfrac{\text{ft}^3}{\text{sec}}\right)\right|\end{array}\right)}$$

$$= 0.17 \text{ ft}^3/\text{sec}$$

Trial 2, with corrected flows of 1.42 ft³/sec in pipe AB and 0.58 ft³/sec in pipes BC and CA,

$$\Delta_2 = \frac{-\left(\begin{array}{c}(63)\left(1.42\,\dfrac{\text{ft}^3}{\text{sec}}\right)^2 - (382)\left(0.58\,\dfrac{\text{ft}^3}{\text{sec}}\right)^2 \\ -(35)\left(0.58\,\dfrac{\text{ft}^3}{\text{sec}}\right)^2\end{array}\right)}{(2)\left(\begin{array}{c}\left|(63)\left(1.42\,\dfrac{\text{ft}^3}{\text{sec}}\right)\right| + \left|(382)\left(-0.58\,\dfrac{\text{ft}^3}{\text{sec}}\right)\right| \\ + \left|(35)\left(-0.58\,\dfrac{\text{ft}^3}{\text{sec}}\right)\right|\end{array}\right)}$$

$$= 0.020 \text{ ft}^3/\text{sec}$$

Trial 3, with corrected flows of 1.44 ft³/sec in pipe AB and 0.56 ft³/sec in pipes BC and CA,

$$\Delta_3 = \frac{-\left(\begin{array}{c}(63)\left(1.44\,\dfrac{\text{ft}^3}{\text{sec}}\right)^2 - (382)\left(0.56\,\dfrac{\text{ft}^3}{\text{sec}}\right)^2 \\ -(35)\left(0.56\,\dfrac{\text{ft}^3}{\text{sec}}\right)^2\end{array}\right)}{(2)\left(\begin{array}{c}\left|(63)\left(1.44\,\dfrac{\text{ft}^3}{\text{sec}}\right)\right| + \left|(382)\left(-0.56\,\dfrac{\text{ft}^3}{\text{sec}}\right)\right| \\ + \left|(35)\left(-0.56\,\dfrac{\text{ft}^3}{\text{sec}}\right)\right|\end{array}\right)}$$

$$= 0.00020 \text{ ft}^3/\text{sec}$$

Accept flows used in trial 3.

For pipe BC,
$$Q = 0.56 \text{ ft}^3/\text{sec}$$

A pipe cross-sectional area ft²

$$A = \pi\frac{D^2}{4} = \frac{\pi(0.33)^2}{4} = 0.086 \text{ ft}^2$$

v flow velocity ft/sec

$$v = \frac{Q}{A} = \frac{0.56\,\dfrac{\text{ft}^3}{\text{sec}}}{0.086 \text{ ft}^2} = 6.5 \text{ ft/sec}$$

The answer is (C).

Why Other Options Are Wrong

(A) This incorrect solution presents the flow rate in velocity units and does not calculate the flow velocity. Other assumptions, definitions, and equations are unchanged from the correct solution.

(B) This incorrect solution uses 0.50 ft diameter in place of 0.33 ft to calculate the flow velocity. Other assumptions, definitions, and equations are unchanged from the correct solution.

(D) This incorrect solution uses the flow rate for pipe AB instead of pipe BC to calculate flow velocity. Other assumptions, definitions, and equations are unchanged from the correct solution.

SOLUTION 29

Typical average water demand for residential use varies between about 75 to 130 gal/person day. These uses are augmented by losses and public and commercial activities that support the community. Because of these augmenting activities, the average annual daily demand is generally taken to be about 160 to 165 gal/person-day. During summer months, the demand may increase by 20% to 30% and daily maximums may add another 50% to 100%. Fire demand contributes to increased water demand, with 500 gal/min as the acceptable minimum. Based on population, fire demand can be estimated by

P population $(10)^3$ people
Q_f fire demand gal/min

$$Q_f = 1020\sqrt{P}\left(1 - 0.01\sqrt{P}\right)$$
$$= 1020\sqrt{3.2}\left(1 - \left(0.01\sqrt{3.2}\right)\right)$$
$$= 1792 \text{ gal/min}$$

Accept the fire demand because 1792 gal/min is greater than 500 gal/min.

f_p daily peak correction factor %
f_s seasonal correction factor %
Q total maximum daily demand gal/min
Q_m average annual daily demand gal/min

Assume that the following are representative.

parameter	value
Q_m	165 gal/person-day
f_s	25% or 1.25
f_p	75% or 1.75

$$Q = Q_m P f_s f_p + Q_f$$
$$= \left(165 \; \frac{\text{gal}}{\text{person-day}}\right)(3200 \text{ people})\left(\frac{1 \text{ day}}{1440 \text{ min}}\right)$$
$$\times (1.25)(1.75) + 1792 \; \frac{\text{gal}}{\text{min}}$$
$$= 2594 \text{ gal/min} \quad (2600 \text{ gal/min})$$

The answer is (D).

Why Other Options Are Wrong

(A) This incorrect solution uses the typical residential demand instead of the annual average demand, does not include the seasonal and daily peak multipliers, and uses the minimum fire demand instead of calculating fire demand on the basis of population. Other assumptions, definitions, and equations are unchanged from the correct solution.

(B) This incorrect solution uses the minimum fire demand and multiplies the sum of the per capita demand and fire demand by the seasonal and daily peak multipliers. Other assumptions, definitions, and equations are unchanged from the correct solution.

(C) This incorrect solution does not include the seasonal and daily peak multipliers. Other assumptions, definitions, and equations are unchanged from the correct solution.

SOLUTION 30

The Safe Drinking Water Act (SDWA) regulates all public drinking water systems in the United States. Public drinking water systems include all those serving at least 15 connections or serving at least 25 people for 60 days of each year. The purpose of the SDWA is to protect public health from exposure to both naturally occurring and man-made contaminants in the water supply. To satisfy this purpose, the SDWA includes regulation of waste disposal by injection into the groundwater.

The answer is (C).

Why Other Options Are Wrong

(A) This choice is incorrect because the SDWA does regulate all public drinking water systems in the United States.

(B) This choice is incorrect because the SDWA does regulate both naturally occurring and man-made contaminants in drinking water.

(D) This choice is incorrect because the SDWA does regulate injection of wastes through injection wells. The regulation occurs under the Underground Injection Control (UIC) program.

SOLUTION 31

time period (hr)	average demand (gal/min)
0000–0200	6900
0200–0400	6500
0400–0600	8100
0600–0800	12,100
0800–1000	14,300
1000–1200	15,600
1200–1400	13,800
1400–1600	12,500
1600–1800	9900
1800–2000	8900
2000–2200	8300
2200–2400	7200
	124,100

n number of time periods
q_m average time period demand   gal/min
Q_m average daily demand   gal/min

$$Q_m = \frac{\sum q_m}{n}$$
$$= \frac{124{,}100 \; \frac{\text{gal}}{\text{min}}}{12}$$
$$= 10{,}342 \text{ gal/min}$$

The reservoir is filling (net flow into the reservoir) during those time periods when the average time period demand is less than the average daily demand.

time period (hr)	average demand (gal/min)	net flow
0000–0200	6900	in
0200–0400	6500	in
0400–0600	8100	in
0600–0800	12,100	out
0800–1000	14,300	out
1000–1200	15,600	out
1200–1400	13,800	out
1400–1600	12,500	out
1600–1800	9900	in
1800–2000	8900	in
2000–2200	8300	in
2200–2400	7200	in

Sometime between 5:00 a.m. and 7:00 a.m. and between 3:00 p.m. and 5:00 p.m., the net flow changes direction. The midpoint of the time interval was used because the average demand for the 2 hr time interval was given. Assume that the flow-time relationship is linear during any 2 hr interval.

For the period from 5:00 a.m. to 7:00 a.m., apply linear interpolation to find that the change occurs at

$$0500 + \frac{(2 \text{ hr})(10{,}342 - 8100)}{12{,}100 - 8100} = 0500 + 1.12 \text{ hr}$$
$$= 0607 \quad (6{:}10 \text{ a.m.})$$

For the period from 1500 to 1700, apply linear interpolation to find that the change occurs at

$$1700 - \frac{(2 \text{ hr})(10{,}342 - 9900)}{12{,}500 - 9900} = 1700 - 0.34 \text{ hr}$$
$$= 1640 \quad (4{:}40 \text{ p.m.})$$

The net flow into the reservoir occurs from approximately 4:40 p.m. to 6:10 a.m.

The answer is (D).

Why Other Options Are Wrong

(A) This incorrect solution gives the time period during which the net flow is out of the reservoir and does not interpolate to define the time that the net flow direction changes. Other assumptions, definitions, and equations are unchanged from the correct solution.

(B) This incorrect solution does not interpolate to define the time that the net flow direction changes. Other assumptions, definitions, and equations are unchanged from the correct solution.

(C) This incorrect solution makes an error in the linear interpolation calculation between 1600 and 1800. Other assumptions, definitions, and equations are unchanged from the correct solution.

Depth Solutions

HYDRAULICS—CLOSED CONDUIT

SOLUTION 32

Assume the inside pipe diameter is equal to the nominal diameter.

D inside pipe diameter   in
ε roughness coefficient for steel pipe   ft
$\dfrac{\varepsilon}{D}$ relative roughness   –

$$\varepsilon = 0.0002 \text{ ft}$$

$$\frac{\varepsilon}{D_6} = \frac{(0.0002 \text{ ft})\left(\dfrac{12 \text{ in}}{\text{ft}}\right)}{6 \text{ in}}$$
$$= 0.0004$$

$$\frac{\varepsilon}{D_4} = \frac{(0.0002 \text{ ft})\left(\dfrac{12 \text{ in}}{\text{ft}}\right)}{4 \text{ in}}$$
$$= 0.0006$$

A pipe cross-sectional area   ft^2

$$A = \pi \frac{D^2}{4}$$

$$A_6 = \frac{\pi (6 \text{ in})^2 \left(\dfrac{1 \text{ ft}^2}{144 \text{ in}^2}\right)}{4}$$
$$= 0.196 \text{ ft}^2$$

$$A_4 = \frac{\pi (4 \text{ in})^2 \left(\dfrac{1 \text{ ft}^2}{144 \text{ in}^2}\right)}{4}$$
$$= 0.087 \text{ ft}^2$$

Re   Reynolds number   –
f friction factor   –

Assume a Reynolds number of 5×10^5 and a temperature of 60°F.

From a Moody diagram, estimate a friction factor using Re and ε/D,

$$f_6 = 0.017$$
$$f_4 = 0.0185$$

g gravitational acceleration   32.2 ft/sec^2
L pipe length   ft
v flow velocity   ft/sec

For parallel pipes, the friction losses must be equal.

$$\frac{f_6 L_6 v_6^2}{2 D_6 g} = \frac{f_4 L_4 v_4^2}{2 D_4 g}$$

$$\frac{(0.017)(1400 \text{ ft})v_6^2}{(2)(6 \text{ in})\left(\dfrac{1 \text{ ft}}{12 \text{ in}}\right) \times \left(32.2 \dfrac{\text{ft}}{\text{sec}^2}\right)} = \frac{(0.0185)(1400 \text{ ft})v_4^2}{(2)(4 \text{ in})\left(\dfrac{1 \text{ ft}}{12 \text{ in}}\right) \times \left(32.2 \dfrac{\text{ft}}{\text{sec}^2}\right)}$$

$$0.74 v_6^2 = 1.21 v_4^2$$
$$v_6 = 1.3 v_4$$

Q total flow rate   ft^3/sec

$$Q = A_6 v_6 + A_4 v_4$$
$$2.4 \frac{\text{ft}^3}{\text{sec}} = (0.196 \text{ ft}^2)(1.3 v_4) + (0.087 \text{ ft}^2)(v_4)$$
$$v_4 = 7.0 \text{ ft/sec}$$
$$v_6 = (1.3)\left(7.0 \frac{\text{ft}}{\text{sec}}\right) = 9.1 \text{ ft/sec}$$

Check the Reynolds number assumption.

ν kinematic viscosity   ft^2/sec

$$\text{Re} = \frac{D v}{\nu}$$

$$\nu = 1.217 \times 10^{-5} \frac{\text{ft}^2}{\text{sec}} \text{ at } 60°\text{F}$$

$$\text{Re}_6 = \frac{(6 \text{ in})\left(\dfrac{1 \text{ ft}}{12 \text{ in}}\right)\left(9.1 \dfrac{\text{ft}}{\text{sec}}\right)}{1.217 \times 10^{-5} \dfrac{\text{ft}^2}{\text{sec}}} = 3.7 \times 10^5$$

$$\text{Re}_4 = \frac{(4 \text{ in})\left(\dfrac{1 \text{ ft}}{12 \text{ in}}\right)\left(7.0 \dfrac{\text{ft}}{\text{sec}}\right)}{1.217 \times 10^{-5} \dfrac{\text{ft}^2}{\text{sec}}} = 1.9 \times 10^5$$

$$f = \frac{0.25}{\left(\log\left(\dfrac{\dfrac{\varepsilon}{D}}{3.7} + \dfrac{5.74}{\text{Re}^{0.9}}\right)\right)^2}$$

44 SIX-MINUTE SOLUTIONS FOR CIVIL PE EXAM PROBLEMS

$$f_6 = \frac{0.25}{\left(\log\left(\frac{0.0004}{3.7} + \frac{5.74}{(3.7 \times 10^5)^{0.9}}\right)\right)^2}$$

$$= 0.0174 \cong 0.017 \quad \text{[close enough]}$$

$$f = \frac{0.25}{\left(\log\left(\frac{0.0006}{3.7} + \frac{5.74}{(1.9 \times 10^5)^{0.9}}\right)\right)^2}$$

$$= 0.0195 \cong 0.0185 \quad \text{[close enough]}$$

$$Q = Av$$

$$Q_4 = (0.087 \text{ ft}^2)\left(7.0 \frac{\text{ft}}{\text{sec}}\right)$$

$$= 0.609 \text{ ft}^3/\text{sec} \quad (0.61 \text{ ft}^3/\text{sec})$$

The answer is (B).

Why Other Options Are Wrong

(A) This incorrect solution fails to square the velocity terms in the friction factor equation. Other assumptions, definitions, and equations are unchanged from the correct solution.

(C) This incorrect solution assumes a relative roughness of 0.002 as a mid-range value on the Moody diagram. Other assumptions, definitions, and equations are unchanged from the correct solution.

(D) This incorrect solution simply proportions the flow based on the area of each pipe. Other assumptions, definitions, and equations are unchanged from the correct solution.

SOLUTION 33

h	elevation head	ft
n	number of stages	–
N_s	specific speed	–
Q	pump discharge	gal/min
ω	rotating speed	rev/min

$$N_s = \frac{\omega\sqrt{Q}}{\left(\frac{h}{n}\right)^{3/4}}$$

$$\left(\frac{h}{n}\right)^{3/4} = \frac{\omega\sqrt{Q}}{N_s}$$

$$n = h\left(\frac{N_s}{\omega\sqrt{Q}}\right)^{4/3}$$

$$= (350 \text{ ft})\left(\frac{2300}{\left(1750 \frac{\text{rev}}{\text{min}}\right)\sqrt{\left(800 \frac{\text{gal}}{\text{min}}\right)}}\right)^{4/3}$$

$$= 5.8 \quad \text{(6 stages)}$$

The answer is (C).

Why Other Options Are Wrong

(A) This choice incorrectly rearranges the specific speed equation. Other definitions and equations are unchanged from the correct solution.

(B) This incorrect choice takes the square root of both the flow and rotational speed in the number of stages equation and fails to multiply by the head, although the equation is correct. Other definitions and equations are unchanged from the correct solution.

(D) This incorrect choice includes a math error in the application of exponents in developing the equation for the number of stages. Other definitions and equations are unchanged from the correct solution.

SOLUTION 34

g	gravitational acceleration	9.81 m/s^2
h_s	suction head	m
p_o	atmospheric pressure at 1370 m above mean sea level	84 kN/m^2
p_v	vapor pressure of water at 13°C	1.5 kN/m^2
TDH	total dynamic head	m
ρ	density of water at 13°C	1000 kg/m^3
σ	cavitation constant	–

$$\rho g = \left(1000 \frac{\text{kg}}{\text{m}^3}\right)\left(9.81 \frac{\text{m}}{\text{s}^2}\right)\left(\frac{\text{N}\cdot\text{s}^2}{1 \text{ kg}\cdot\text{m}}\right)\left(\frac{1 \text{ kN}}{1000 \text{ N}}\right)$$

$$= 9.81 \text{ kN/m}^3$$

$$h_s = \sigma\text{TDH} - \frac{p_o}{\rho g} + \frac{p_v}{\rho g}$$

$$= (0.26)(38 \text{ m}) - \frac{84 \frac{\text{kN}}{\text{m}^2}}{9.81 \frac{\text{kN}}{\text{m}^3}} + \frac{1.5 \frac{\text{kN}}{\text{m}^2}}{9.81 \frac{\text{kN}}{\text{m}^3}}$$

$$= 1.47 \text{ m} \quad (1.5 \text{ m})$$

The answer is (B).

Why Other Options Are Wrong

(A) This incorrect solution adds instead of subtracts the vapor pressure term in the suction head equation. Other definitions and equations are unchanged from the correct solution.

(C) This incorrect solution uses the customary U.S. value for specific weight for water in SI units instead of a density-gravitational acceleration term. Other definitions and equations are unchanged from the correct solution.

(D) This incorrect solution transposes the cavitation constant value. Other definitions and equations are unchanged from the correct solution.

SOLUTION 35

fitting	quantity	unit effective length (ft)	total effective length (ft)
square inlet	1	47	47
gate valve	10	3.2	32
standard radius 90° ell	19	21	399
standard radius 45° ell	37	15	555
straight tee	8	7.2	58
			1091

L total pipe length m
L_e effective pipe length m
L_p actual pipe length m

$$L = L_p + L_e$$
$$= (5 \text{ km}) \left(1000 \frac{\text{m}}{\text{km}}\right) + (1091 \text{ ft}) \left(\frac{1 \text{ m}}{3.28 \text{ ft}}\right)$$
$$= 5333 \text{ m}$$

D pipe diameter m

$$D = (400 \text{ mm}) \left(\frac{1 \text{ m}}{1000 \text{ mm}}\right)$$
$$= 0.4 \text{ m}$$

A pipe cross-sectional area m^2

$$A = \pi \frac{D^2}{4}$$
$$= \frac{\pi (0.4 \text{ m})^2}{4}$$
$$= 0.13 \text{ m}^2$$

Assume a Hazen-Williams coefficient of 130 for new welded steel pipe.

C Hazen-Williams coefficient –
h_f head loss from friction (include minor losses as effective pipe length) m
Q flow rate m^3/s

$$h_f = \frac{10.7 Q^{1.85} L}{C^{1.85} D^{4.87}}$$
$$= \frac{(10.7) Q^{1.85} (5333 \text{ m})}{(130)^{1.85} (0.4 \text{ m})^{4.87}}$$
$$= 607 Q^{1.85}$$

g gravitational acceleration 9.81 m/s^2
p pressure N/m^2
z elevation m
ρ water density 1000 kg/m^3

$$\frac{p_1}{\rho g} + z_1 + \frac{Q_1^2}{2g A_1^2} = \frac{p_2}{\rho g} + z_2 + \frac{Q_2^2}{2g A_2^2} + h_f$$

Because the pipeline connects two reservoirs, assume the pressures at the reservoirs are equal and cancel. Assume the flow at the upstream reservoir to be zero.

$$0 + 1100 \text{ m} + 0 = 0 + 835 \text{ m}$$
$$+ \frac{Q_2^2}{(2)\left(9.81 \frac{\text{m}}{\text{s}^2}\right)(0.13 \text{ m}^2)^2}$$
$$+ 607 Q^{1.85}$$
$$1100 \text{ m} - 835 \text{ m} = 3.0 Q^2 + 607 Q^{1.85}$$

Assume the velocity head of $3.0 Q^2$ is negligible compared to the friction loss of $607 Q^{1.85}$.

$$265 = 607 Q^{1.85}$$
$$Q = \left(\frac{265}{607}\right)^{1/1.85}$$
$$= 0.64 \text{ m}^3/\text{s}$$

The answer is (C).

Why Other Options Are Wrong

(A) This incorrect solution uses a Hazen-Williams coefficient of 100 instead of 130. For new pipe, the typical value is 130. Other assumptions, definitions, and equations are the same as used in the correct solution.

(B) This incorrect solution ignores minor losses and uses a Hazen-Williams coefficient of 100. Other assumptions, definitions, and equations are the same as used in the correct solution.

(D) This incorrect solution ignores minor losses and uses a wrong Hazen-Williams coefficient. Other assumptions, definitions, and equations are the same as used in the correct solution.

SOLUTION 36

β	beta ratio	–
D_1	upstream pipe diameter	in or ft
D_2	throat diameter	in or ft

$$\beta = \frac{D_2}{D_1} = \frac{10}{16} = 0.625$$

C_v	velocity coefficient	–
g	gravitational acceleration	32.2 ft/sec^2
h	manometer reading	ft
ρ	water density	62.3 lbm/ft^3 at 70°F
ρ_m	manometer fluid density	848 lbm/ft^3 at 70°F
v_2	throat velocity	ft/sec

Assume that the velocity coefficient is 1.0.

$$v_2 = \left(\frac{C_v}{\sqrt{1-\beta^4}}\right)\sqrt{\left(\frac{2g}{\rho}\right)(\rho_m - \rho)h}$$

$$= \left(\frac{1}{\sqrt{1-0.625^4}}\right)$$

$$\times \sqrt{\frac{(2)\left(32.2 \frac{\text{ft}}{\text{sec}^2}\right) \times \left(848 \frac{\text{lbm}}{\text{ft}^3} - 62.3 \frac{\text{lbm}}{\text{ft}^3}\right)(9.4 \text{ in})}{\left(62.3 \frac{\text{lbm}}{\text{ft}^3}\right)\left(\frac{12 \text{ in}}{\text{ft}}\right)}}$$

$$= 27.4 \text{ ft/sec}$$

A_2 throat cross-sectional area   ft^2

$$A_2 = \pi \frac{D_2^2}{4} = \left(\frac{\pi (10 \text{ in})^2}{4}\right)\left(\frac{1 \text{ ft}^2}{144 \text{ in}^2}\right)$$

$$= 0.55 \text{ ft}^2$$

Q flow rate   ft^3/sec

$$Q = A_2 v_2 = (0.55 \text{ ft}^2)\left(27.4 \frac{\text{ft}}{\text{sec}}\right)$$

$$= 15 \text{ ft}^3/\text{sec}$$

The answer is (B).

Why Other Options Are Wrong

(A) This incorrect solution inverts the beta ratio. Other assumptions, definitions, and equations are unchanged from the correct solution.

(C) This incorrect solution fails to include the exponent in the beta ratio term in the velocity equation. Other assumptions, definitions, and equations are unchanged from the correct solution.

(D) This incorrect solution applies the area of the pipe instead of the area of the throat in the velocity equation. Other assumptions, definitions, and equations are unchanged from the correct solution.

HYDRAULICS—OPEN CHANNEL

SOLUTION 37

g	gravitational acceleration	32.2 ft/sec^2
h	head	ft
v	velocity	ft/sec

$$v = \sqrt{2gh} = \sqrt{(2)\left(32.2 \frac{\text{ft}}{\text{sec}^2}\right)(42 \text{ ft})}$$

$$= 52 \text{ ft/sec}$$

A_c	chute cross-sectional area	ft^2
Q	flow	ft^3/sec

$$A_c = \frac{Q}{v} = \frac{200 \frac{\text{ft}^3}{\text{sec}}}{52 \frac{\text{ft}}{\text{sec}}}$$

$$= 3.8 \text{ ft}^2$$

D_c	chute flow depth	ft
w_c	chute width	ft

$$D_c = \frac{A_c}{w_c} = \frac{3.8 \text{ ft}^2}{3 \text{ ft}}$$

$$= 1.267 \text{ ft}$$

Fr   Froude number   –

$$\text{Fr} = \frac{v}{\sqrt{gD_c}} = \frac{52 \frac{\text{ft}}{\text{sec}}}{\sqrt{\left(32.2 \frac{\text{ft}}{\text{sec}^2}\right)(1.267 \text{ ft})}}$$

$$= 8.14$$

w_b basin width   ft

$$\frac{w_b}{D_c} = 0.875 \text{ Fr} + 2.7$$

$$w_b = D_c(0.875 \text{ Fr} + 2.7)$$

$$= (1.267 \text{ ft})\big((0.875)(8.14) + 2.7\big)$$

$$= 12.44 \text{ ft} \quad (12 \text{ ft})$$

The answer is (C).

Why Other Options Are Wrong

(A) This incorrect solution assumes the depth of flow and minimum required basin width to be equal. Other definitions and equations are unchanged from the correct solution.

(B) This incorrect solution inverts the values for area and flow in the area equation. Other definitions and equations are unchanged from the correct solution.

(D) This incorrect solution takes the square root of the head instead of taking the square root of the head and gravity constant term. Other definitions and equations are unchanged from the correct solution.

SOLUTION 38

Assume uniform flow since flow velocity is relatively slow and slope is constant and shallow.

D pipe diameter   m
R hydraulic radius   m
θ angle of flow cross section   radians

For flow half full in a circular culvert, θ is π radians.

$$R = 0.25\left(1 - \frac{\sin\theta}{\theta}\right)D$$
$$= 0.25\left(1 - \frac{\sin\pi}{\pi}\right)D$$
$$= 0.25D$$

d_2 flow depth in culvert   m

$$d_2 = \frac{D}{2}$$

Standard concrete pipe sizes go up to 9 ft (2.74 m). When the culvert flows half full,

$$d_2 = \frac{2.74 \text{ m}}{2} = 1.37 \text{ m}$$

n Manning roughness coefficient   –
S culvert slope   m/m
v_2 flow velocity in culvert   m/s

For concrete pipe, the Manning roughness coefficient is 0.013.

$$v_2 = \frac{R^{2/3}\sqrt{S}}{n}$$
$$= \frac{\left((0.25)(2)(1.37)\right)^{2/3}\sqrt{0.002 \frac{\text{m}}{\text{m}}}}{0.013}$$
$$= 2.67 \text{ m/s}$$

A_1 channel cross-sectional area   m²
A_2 culvert cross-sectional area   m²
v_1 flow velocity in channel   m/s

Apply the continuity equation.

$$A_1v_1 = A_2v_2$$
$$A_1v_1 = (1.5 \text{ m})(8 \text{ m})\left(2.5\frac{\text{m}}{\text{s}}\right)$$
$$= 30 \text{ m}^3/\text{s}$$
$$A_2 = \left(\frac{1}{2}\right)\left(\frac{\pi D^2}{4}\right)$$
$$= \left(\frac{1}{2}\right)(2.74 \text{ m}^2)$$
$$= 2.95 \text{ m}^2$$

Try one pipe.

$$A_2v_2 = \left(2.95\frac{\text{m}^2}{\text{pipe}}\right)\left(2.67\frac{\text{m}}{\text{s}}\right)$$
$$= 7.88 \text{ m}^3/\text{s per pipe}$$

$$\text{number of 9 ft diameter pipes} = \frac{30\frac{\text{m}^3}{\text{s}}}{7.88\frac{\frac{\text{m}^3}{\text{s}}}{\text{pipe}}}$$
$$= 3.8 \text{ pipes}$$

Use four pipes, each with a diameter of 9 ft.

The answer is (C).

Why Other Options Are Wrong

(A) This incorrect solution uses the total pipe cross-sectional area instead of the area of flow when applying the continuity equation. Other assumptions, definitions, and equations are unchanged from the correct solution.

(B) This incorrect solution uses a different approach and makes an error in combining the exponents in the continuity equation. Other assumptions, definitions, and equations are unchanged from the correct solution.

(D) This incorrect solution uses a different approach and uses $\pi/2$ instead of π for the angle of cross-sectional flow. Other assumptions, definitions, and equations are unchanged from the correct solution.

SOLUTION 39

Behind the dam,

d_1 water depth behind the dam m
E_1 energy line behind the dam m

$$E_1 = d_1 = 21 \text{ m} + 2 \text{ m}$$
$$= 23 \text{ m}$$

At the spillway crest, the critical energy is equal to the water depth above the crest.

d_c critical depth at the spillway crest m
E_c energy line at the spillway crest m

$$E_c = 2 \text{ m}$$
$$d_c = \left(\frac{2}{3}\right) E_c = \left(\frac{2}{3}\right) (2 \text{ m})$$
$$= 1.33 \text{ m}$$

g gravitational acceleration 9.81 m/s²
v_c critical water velocity m/s

$$E_c = d_c + \frac{v_c^2}{2g}$$
$$2 \text{ m} = 1.33 \text{ m} + \frac{v_c^2}{(2)\left(9.81 \frac{\text{m}}{\text{s}^2}\right)}$$
$$v_c = 3.62 \text{ m/s}$$

q flow rate per unit width m²/s

$$q = v_c d_c = \left(3.62 \frac{\text{m}}{\text{s}}\right)(1.33 \text{ m})$$
$$= 4.81 \text{ m}^2/\text{s}$$

At the toe of the dam,

d_2 water depth at the toe of the dam m
E_2 energy line at the toe of the dam m

$$E_1 = E_2 = d_2 + \frac{q^2}{2gd_2^2}$$
$$23 \text{ m} = d_2 + \frac{\left(4.81 \frac{\text{m}^2}{\text{s}}\right)^2}{(2)\left(9.81 \frac{\text{m}}{\text{s}^2}\right) d_2^2} = d_2 + \frac{1.18}{d_2^2}$$

Solve for d_2 by trial and error.

$$d_2 = 0.23 \text{ m}$$

F_d force on the dam per unit width kN/m
ρ density of water 1000 kg/m³

$$F_d = \rho g \left(\frac{d_1^2}{2} + \frac{q^2}{gd_1} - \frac{d_2^2}{2} - \frac{q^2}{gd_2}\right)$$

$$\rho g = \left(1000 \frac{\text{kg}}{\text{m}^3}\right)\left(9.81 \frac{\text{m}}{\text{s}^2}\right)\left(\frac{\text{N·s}^2}{\text{kg·m}}\right)\left(\frac{1 \text{ kN}}{10^3 \text{ N}}\right)$$
$$= 9.81 \text{ kN/m}^3$$

$$F_d = \left(9.81 \frac{\text{kN}}{\text{m}^3}\right)$$
$$\times \left(\frac{(23 \text{ m})^2}{2} + \frac{\left(4.81 \frac{\text{m}^2}{\text{s}}\right)^2}{\left(9.81 \frac{\text{m}}{\text{s}^2}\right)(23 \text{ m})} - \frac{(0.23 \text{ m})^2}{2} - \frac{\left(4.81 \frac{\text{m}^2}{\text{s}}\right)^2}{\left(9.81 \frac{\text{m}}{\text{s}^2}\right)(0.23 \text{ m})}\right)$$

$$= 2495 \text{ kN/m} \quad (2500 \text{ kN/m})$$

The answer is (D).

Why Other Options Are Wrong

(A) This incorrect solution assumes that at every point the energy line is equal to the depth of the water above the crest and that this is also equal to the critical depth. Other assumptions, definitions, and equations are unchanged from the correct solution.

(B) This incorrect solution fails to square the depth terms in the energy equation at the toe of the dam and in the force equation. Other assumptions, definitions, and equations are unchanged from the correct solution.

(C) This incorrect solution uses the critical depth as the depth of the water above the spillway. Other assumptions, definitions, and equations are unchanged from the correct solution.

SOLUTION 40

For a slope of 5% with 20 cm diameter logs placed at 4.0 m intervals, the conditions depicted in the following illustration will result, and the maximum water velocity will occur as the water flows over each log.

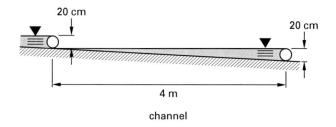

channel

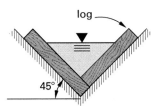

channel cross section

A	flow area through the V-notch formed by the logs	m²
d	water depth over the log	m
θ	side slope angle measure from the horizontal	degree

$$A = \frac{d^2}{\tan \theta} = \frac{d^2}{\tan 45°} = d^2$$

Q	flow rate	m³/s
v	allowable scour velocity	m/s

With the logs placed perpendicular to the flow in a 90° V-shaped cross section channel, they will act as 90° V-notch weirs, and the flow over the weir can be approximated by

$$Q = A\text{v} = 1.4 d^{2.5}$$

$$d^2 \left(0.76 \; \frac{\text{m}}{\text{s}}\right) = 1.4 d^{2.5}$$

$$\sqrt{d} = \frac{0.76 \; \frac{\text{m}}{\text{s}}}{1.4} = 0.54$$

$$d = 0.29 \text{ m}$$

$$A = (0.29 \text{ m})^2 = 0.084 \text{ m}^2$$

$$Q = (0.84 \text{ m}^2) \left(0.76 \; \frac{\text{m}}{\text{s}}\right) = 0.064 \text{ m}^3/\text{s}$$

The answer is (C).

Why Other Options Are Wrong

(A) This incorrect solution ignores the influence of the logs. Other assumptions, definitions, and equations are the same as used in the correct solution.

(B) This incorrect solution ignores the influence of the logs and does not square the depth term in the area calculation. Other assumptions, definitions, and equations are the same as used in the correct solution.

(D) This incorrect solution makes an error in the calculation of the flow depth through the logs. Other assumptions, definitions, and equations, and the figure are the same as used in the correct solution.

SOLUTION 41

m	mass of chemical released	lbm
SG	chemical specific gravity	–
V	release volume	ft³
ρ_c	chemical density	–
ρ_w	water density	lbm/ft³

$$m = \rho_c V = \text{SG} \rho_w V$$

$$= (0.9)\left(62.4 \; \frac{\text{lbm}}{\text{ft}^3}\right)(200 \text{ gal})\left(0.134 \; \frac{\text{ft}^3}{\text{gal}}\right)$$

$$= 1505 \text{ lbm}$$

Assume the ditch cross section is rectangular along its entire length.

A	ditch wetted cross-sectional area	ft²
d	water depth	ft
w	ditch width	ft

$$A = wd = (6 \text{ ft})(2 \text{ ft})$$
$$= 12 \text{ ft}^2$$

g	gravitational constant	32.2 ft/sec²
s	ditch slope	ft/ft
v*	shear velocity	ft/day

$$\text{v}^* = \sqrt{gds} = \sqrt{\left(32.2 \; \frac{\text{ft}}{\text{sec}^2}\right)(2 \text{ ft})\left(\frac{1 \text{ ft}}{100 \text{ ft}}\right)}$$
$$= 0.80 \text{ ft/sec}$$

E_L	longitudinal dispersion coefficient	ft²/day
v	ditch flow average velocity	ft/day

$$E_L = \frac{0.011 \text{v}^2 w^2}{d\text{v}^*}$$

$$= \frac{(0.011)\left(3 \; \frac{\text{ft}}{\text{sec}}\right)^2 (6 \text{ ft})^2 \left(86{,}400 \; \frac{\text{sec}}{\text{day}}\right)}{(2 \text{ ft})\left(0.80 \; \frac{\text{ft}}{\text{sec}}\right)}$$

$$= 192{,}456 \text{ ft}^2/\text{day}$$

t	time	day
x	downstream distance of travel	ft

$$t = \frac{x}{\text{v}} = \frac{(1.8 \text{ mi})\left(5280 \; \frac{\text{ft}}{\text{mi}}\right)}{\left(3 \; \frac{\text{ft}}{\text{sec}}\right)\left(86{,}400 \; \frac{\text{sec}}{\text{day}}\right)} = 0.037 \text{ day}$$

$C_{\max}$	maximum concentration	mg/L
K	ditch dispersion coefficient	day⁻¹

$$c_{max} = \frac{Me^{-Kt}}{A\sqrt{4\pi E_L t}}$$

$$= \frac{(1505 \text{ lbm})\left(e^{(-0.3/\text{day})(0.037 \text{ day})}\right) \times \left(\frac{10^6 \text{ mg}}{2.204 \text{ lbm}}\right)\left(0.0354 \frac{\text{ft}^3}{\text{L}}\right)}{(12 \text{ ft}^2)\sqrt{4\pi\left(192{,}456 \frac{\text{ft}^2}{\text{day}}\right)(0.037 \text{ day})}}$$

$$= 6660 \text{ mg/L} \quad (6700 \text{ mg/L})$$

The answer is (B).

Why Other Options Are Wrong

(A) This incorrect solution neglects to take the square root of shear velocity. Other definitions, assumptions, and equations are the same as used in the correct solution.

(C) This incorrect solution neglects to take the negative power of e in the maximum concentration equation. Other definitions, assumptions, and equations are the same as used in the correct solution.

(D) This incorrect solution uses the conversion for kilometers to feet instead of miles to feet in the time equation. Other definitions, assumptions, and equations are the same as used in the correct solution.

HYDROLOGY

SOLUTION 42

d_e depth of irrigation water applied per plot   cm/h·plot
d_T total annual depth of irrigation water applied   cm/yr
t_p total weekly irrigation time per plot   h/wk·plot

$$t_p = \frac{d_T}{d_e} = \frac{260 \frac{\text{cm}}{\text{yr}}}{\left(2 \frac{\text{cm}}{\text{h·plot}}\right)\left(52 \frac{\text{wk}}{\text{yr}}\right)}$$

$$= 2.5 \text{ h/wk·plot}$$

t_T total available weekly irrigation time   h/wk

$$t_T = \left(12 \frac{\text{h}}{\text{d}}\right)\left(7 \frac{\text{d}}{\text{wk}}\right)$$

$$= 84 \text{ h/wk}$$

n number of plots

$$n = \frac{t_T}{t_p} = \frac{84 \frac{\text{h}}{\text{wk}}}{2.5 \frac{\text{h}}{\text{wk·plot}}}$$

$$= 34 \text{ plots}$$

A_p plot irrigated area   ha
A_T total irrigated area   ha

$$A_p = \frac{A_T}{n} = \frac{500 \text{ ha}}{34 \text{ plots}}$$

$$= 14.7 \text{ ha} \quad (15 \text{ ha})$$

The answer is (B).

Why Other Options Are Wrong

(A) This incorrect solution includes the three irrigation periods in the denominator of the weekly irrigation time equation. Other assumptions, definitions, and equations are unchanged from the correct solution.

(C) This incorrect solution uses a 5 day week instead of a 7 day week. Other assumptions, definitions, and equations are unchanged from the correct solution.

(D) This incorrect solution includes the three irrigation periods in the numerator of the weekly irrigation time equation. Other assumptions, definitions, and equations are unchanged from the correct solution.

SOLUTION 43

At the spillway crest, the critical energy is equal to the water depth above the crest.

E_c energy line at the spillway crest   ft

$$E_c = 6 \text{ ft}$$

d_c critical depth at the spillway crest   ft

$$d_c = \left(\frac{2}{3}\right)E_c = \left(\frac{2}{3}\right)(6 \text{ ft}) = 4 \text{ ft}$$

g gravitational acceleration   32.2 ft/sec^2
v_c critical water velocity   ft/s

$$E_c = d_c + \frac{v_c^2}{2g}$$

$$6 \text{ ft} = 4 \text{ ft} + \frac{v_c^2}{(2)\left(32.2 \frac{\text{ft}}{\text{sec}^2}\right)}$$

$$v_c = 11.3 \text{ ft/sec}$$

Δt the duration of the period of interest   sec

$$\Delta t = (8 \text{ hr})\left(3600 \; \frac{\text{sec}}{\text{hr}}\right) = 28{,}800 \text{ sec}$$

O_1 outflow at the beginning of the
period of interest ft^3/sec

Assume that during dry weather conditions, outflow is negligible.

O_2 outflow at the end of the period
of interest ft^3/sec
w spillway width   ft

$$O_2 = v_c d_c w = \left(11.3 \; \frac{\text{ft}}{\text{sec}}\right)(4 \text{ ft})(30 \text{ ft})$$
$$= 1356 \; \text{ft}^3/\text{sec}$$

I_1 inflow at beginning of the
period of interest ft^3/sec
I_2 inflow at the end of the period
of interest ft^3/sec
Δs the increase in reservoir storage
during the period of interest ft^3

$$\Delta s = \left(\frac{\Delta t}{2}\right)(I_1 + I_2 - O_1 - O_2)$$
$$= \left(\frac{28{,}800 \text{ sec}}{2}\right)\begin{pmatrix} 1100 \; \frac{\text{ft}^3}{\text{sec}} + 1550 \; \frac{\text{ft}^3}{\text{sec}} \\ - 0 - 1356 \; \frac{\text{ft}^3}{\text{sec}} \end{pmatrix}$$
$$\times \left(\frac{1 \text{ ac-ft}}{43{,}560 \text{ ft}^3}\right)$$
$$= 428 \text{ ac-ft} \quad (430 \text{ ac-ft})$$

The answer is (B).

Why Other Options Are Wrong

(A) This incorrect solution uses the rise in the water level above the spillway instead of the critical depth to calculate the outflow from the reservoir during the 10-year storm. Other assumptions, definitions, and equations are unchanged from the correct solution.

(C) This incorrect solution fails to divide by two in the velocity equation. Other assumptions, definitions, and equations are unchanged from the correct solution.

(D) This incorrect solution calculates the total storage volume during the 10-year storm instead of the increase in volume only. Other assumptions, definitions, and equations are unchanged from the correct solution.

SOLUTION 44

d depth of rainfall   cm
i intensity   cm/h
t storm duration   h

From the illustration, for a 50-year recurrence interval and 2 h duration storm, the rainfall depth is 9 cm.

$$i = \frac{d}{t} = \frac{9 \text{ cm}}{2 \text{ h}} = 4.5 \text{ cm/h}$$

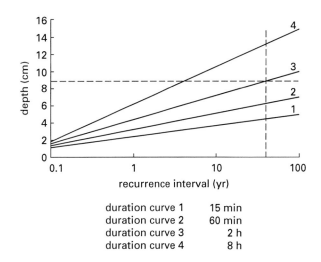

duration curve 1   15 min
duration curve 2   60 min
duration curve 3   2 h
duration curve 4   8 h

For the Steel formula, intensity must be in units of in/hr.

$$i = \left(4.5 \; \frac{\text{cm}}{\text{h}}\right)\left(\frac{1 \text{ in}}{2.54 \text{ cm}}\right) = 1.8 \text{ in/hr}$$

b constant   min
K constant   in-min/hr
t_c time of concentration   min

The Steel formula is

$$i = \frac{K}{t_c + b}$$
$$t_c = \frac{K}{i} - b$$
$$= \frac{180 \; \frac{\text{in-min}}{\text{hr}}}{1.8 \; \frac{\text{in}}{\text{hr}}} - 25 \text{ min}$$
$$= 75 \text{ min}$$

The answer is (D).

Why Other Options Are Wrong

(A) This incorrect solution uses SI units for all calculations instead of converting to customary U.S. units for the Steel formula. The illustration and other definitions and equations are unchanged from the correct solution.

(B) This incorrect solution takes the rainfall as the intensity instead of calculating intensity from rainfall and duration. The illustration is unchanged from the correct solution.

(C) This incorrect solution misreads the illustration, mistaking the 8 h duration for the 2 h duration storm. Other definitions and equations are unchanged from the correct solution.

SOLUTION 45

i rainfall intensity   in/hr
L overland flow distance   ft

$$iL = \left(0.89 \, \frac{\text{in}}{\text{hr}}\right)(150 \text{ ft}) = 134 < 500$$

Because the site is characterized by sheet flow and the product of rainfall intensity in units of in/hr and the flow distance in units of ft is less than 500, the Izzard method for calculating time to concentration applies. The rainfall data is not provided in a format that will allow use of the Manning kinematic equation.

c overland flow retardance coefficient   –
k Izzard coefficient   –
S ground surface slope   ft/ft

For smooth asphalt, the value of the overland flow retardance coefficient is 0.007.

$$k = \frac{0.0007i + c}{S^{1/3}}$$

$$k_A = \frac{(0.0007)\left(0.89 \, \frac{\text{in}}{\text{hr}}\right) + 0.007}{(0.001)^{1/3}} = 0.076$$

$$k_B = \frac{(0.0007)\left(0.89 \, \frac{\text{in}}{\text{hr}}\right) + 0.007}{(0.0006)^{1/3}} = 0.090$$

t_c time to concentration   min

$$t_c = \frac{41kL^{1/3}}{i^{2/3}}$$

$$t_{c,A} = \frac{(41)(0.076)(150 \text{ ft})^{1/3}}{(0.89)^{2/3} \, \frac{\text{in}}{\text{hr}}} = 17.9 \text{ min}$$

$$t_{c,B} = \frac{(41)(0.090)(90 \text{ ft})^{1/3}}{(0.89)^{2/3} \, \frac{\text{in}}{\text{hr}}} = 17.9 \text{ min}$$

$$t_{c,A} = t_{c,B} = 17.9 \text{ min} \quad (18 \text{ min})$$

The answer is (B).

Why Other Options Are Wrong

(A) This incorrect solution uses the Manning kinematic equation and manipulates the rainfall intensity for application to the equation. The Manning equation stipulates specific conditions for rainfall. Other assumptions, definitions, and equations are the same as used in the correct solution.

(C) This incorrect solution uses the Manning roughness coefficient for the overland flow retardance coefficient. Other assumptions, definitions, and equations are the same as used in the correct solution.

(D) This incorrect solution adds the time to concentration for the two lots. Other assumptions, definitions, and equations are the same as used in the correct solution.

SOLUTION 46

The hydrograph for each storm is produced by multiplying the unit hydrograph by the rainfall for each storm. The peak discharge is determined by offsetting the hydrographs by 3 h and then adding them. This is shown in the following table and illustration.

storm discharge (m^3/s)	storm runoff (cm) 1.6	3.1	2.7	combined discharge (m^3/s)	duration (h)
0	0	–	–	0	0
25	40	–	–	40	1
125	200	–	–	200	2
368	588.8	0	–	588.8	3
445	712	77.5	–	789.5	4
425	680	387.5	–	1067.5	5
300	480	1140.8	0	1620.8	6
220	352	1379.5	67.5	1799	7
159	254.4	1317.5	337.5	1909.4	8
118	188.8	930	993.6	2112.4	9
84	134.4	682	1201.5	2017.9	10
60	96	492.9	1147.5	1736.4	11
35	56	365.8	810	1231.8	12
22	35.2	260.4	594	889.6	13
14	22.4	186	429.3	637.7	14
11	17.6	108.5	318.6	444.7	15
8	12.8	68.2	226.8	307.8	16
4	6.4	43.4	162	211.8	17
–	–	34.1	94.5	128.6	18
–	–	24.8	59.4	84.2	19
–	–	12.4	37.8	50.2	20
–	–	–	29.7	29.7	21
–	–	–	21.6	21.6	22
–	–	–	10.8	10.8	23

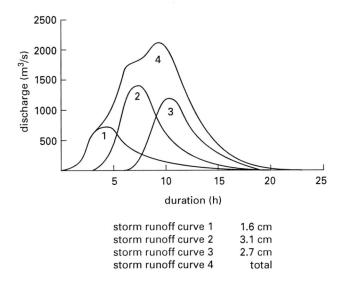

storm runoff curve 1	1.6 cm
storm runoff curve 2	3.1 cm
storm runoff curve 3	2.7 cm
storm runoff curve 4	total

From the illustration, the peak discharge occurs at 9 h after the beginning of the first storm and is equal to 2100 m³/s.

The answer is (B).

Why Other Options Are Wrong

(A) This incorrect solution calculates the hydrograph for the largest storm (3.1 cm) only and then takes the peak for that storm for the peak discharge.

(C) This incorrect solution adds the rainfall for the three storms, multiplies it by the area of the drainage area, and then divides by the storm duration of 3 h. The unit hydrograph is ignored.

(D) This incorrect solution does not offset the hydrographs before adding them to produce the peak hydrograph.

SOLUTION 47

The wilting point is defined as the moisture content of the soil corresponding to a moisture tension of 15 atm during the drying cycle. From the illustration, the moisture content corresponding to a moisture tension of 15 atm is 23%.

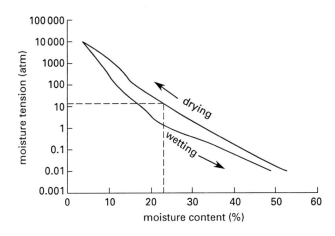

The answer is (C).

Why Other Options Are Wrong

(A) This incorrect solution determined the wilting point using the wetting cycle part of the curve.

(B) This incorrect solution determined the wilting point as the average moisture content between the wetting and drying cycles.

(D) This incorrect solution misread the plot, taking the moisture tension at 1.5 atm instead of 15 atm. The wilting point is 32%.

SOLUTION 48

f	relative humidity	%
p_{a2}	water vapor pressure in air at specified temperature 2 m above lake surface	millibar
p_{s0}	water saturation vapor pressure in air at lake surface temperature	millibar
p_{s2}	water saturation vapor pressure in air at specified temperature 2 m above lake surface	millibar

The saturation vapor pressure of air at 18°C is 20.63 millibar and at 24°C is 29.83 millibar.

$$p_{a2} = \frac{fp_{s2}}{100}$$
$$= \frac{(16\%)(29.83 \text{ millibar})}{100\%}$$
$$= 4.77 \text{ millibar}$$

E	evaporation	mm/d
v_4	wind speed at 4 m above the water surface	m/s

The information provided suggests one of the empirical formulas available that are based on vapor pressure and wind speed. One of these that has been applied to Lake Mead and may be considered to be generally applicable, especially in the southwestern United States, is

$$E = 0.122 \, (p_{s0} - p_{a2}) \, v_4$$
$$= (0.122)(20.63 \text{ millibar} - 4.77 \text{ millibar})\left(3.6 \, \frac{\text{m}}{\text{s}}\right)$$
$$= 7.0 \text{ mm/d}$$

The answer is (D).

Why Other Options Are Wrong

(A) This incorrect solution uses the evaporation equation for customary U.S. units instead of SI units. Other assumptions, definitions, and equations are unchanged from the correct solution.

(B) This incorrect solution uses the saturation vapor pressure for both vapor pressure entries. Other assumptions, definitions, and equations are unchanged from the correct solution.

(C) This incorrect solution uses the vapor pressures in units of mm Hg instead of millibars. Other assumptions, definitions, and equations are unchanged from the correct solution.

SOLUTION 49

L_s total sediment load   lbm/day

From the figure for a stream flow of 20 ft³/sec, the sediment load is about

$$L_s = 100\,000 \text{ lbm/day}$$

V reservoir original volume   ft³
V_s sediment volume   ft³
x reservoir volume fraction occupied by sediment

$$V_s = Vx = (3600 \text{ ac-ft})(0.25)\left(43{,}560 \, \frac{\text{ft}^3}{\text{ac-ft}}\right)$$
$$= 3.9 \times 10^7 \text{ ft}^3$$

Assume typical specific weight values of sediment continuously submerged for 50 years are 50 lbm/ft³ for clay and 65 lbm/ft³ for silt.

f_i fraction of soil type in sediment
γ_i sediment specific weight for soil type fraction   lbm/ft³
γ_s total sediment specific weight at time t lbm/ft³

$$\gamma_s = \sum (f_i \gamma_i) = (0.36)\left(65 \, \frac{\text{lbm}}{\text{ft}^3}\right) + (0.64)\left(50 \, \frac{\text{lbm}}{\text{ft}^3}\right)$$
$$= 55.4 \text{ lbm/ft}^3$$

t reservoir useful life   yr

$$t = \frac{V_s \gamma_s}{L_s} = \frac{(3.9 \times 10^7 \text{ ft}^3)\left(55.4 \, \frac{\text{lbm}}{\text{ft}^3}\right)}{\left(100{,}000 \, \frac{\text{lbm}}{\text{day}}\right)\left(365 \, \frac{\text{day}}{\text{yr}}\right)}$$
$$= 59 \text{ yr}$$

The answer is (B).

Why Other Options Are Wrong

(A) This incorrect solution uses the wrong conversion factor from ft³ to ac-ft in the sediment volume calculation. Other assumptions, definitions, and equations are unchanged from the correct solution.

(C) This incorrect solution does not correct the reservoir volume for the fraction occupied by sediment. Other assumptions, definitions, and equations are unchanged from the correct solution.

(D) This incorrect solution divides by instead of multiplying by the fraction of reservoir volume occupied by sediment. Other assumptions, definitions, and equations are unchanged from the correct solution.

SOLUTION 50

s reservoir storage volume   m³
z water surface elevation   m

$$s = 9.4 z^2 + 3854 z + 14\,470$$

inflow (m³/s)	outflow (m³/s)	water surface elevation (m)	reservoir volume (10⁶ m³)
0.67	0.67	371.1	2.74
1.2	0.70	373.2	2.76
2.8	2.9	379.1	2.83
2.5	3.8	377.8	2.81

I inflow   m³/s
O outflow   m³/s
Δt elapsed time increment   s
$1, 2$ subscripts   successive time steps

For discharge from deep uncontrolled reservoirs, the following equation applies.

$$I_1 + I_2 + \frac{2s_1}{\Delta t} - O_1 = \frac{2s_2}{\Delta t} + O_2$$

$$\Delta t = \frac{2(s_2 - s_1)}{I_1 + I_2 - O_1 - O_2}$$

$$\Delta t_1 = \frac{(2)(2.76 \times 10^6 \text{ m}^3 - 2.74 \times 10^6 \text{ m}^3)\left(\frac{1 \text{ h}}{3600 \text{ s}}\right)}{0.67 \frac{\text{m}^3}{\text{s}} + 1.2 \frac{\text{m}^3}{\text{s}} - 0.67 \frac{\text{m}^3}{\text{s}} - 0.70 \frac{\text{m}^3}{\text{s}}}$$
$$= 22 \text{ h}$$

$$\Delta t_2 = \frac{(2)(2.83 \times 10^6 \text{ m}^3 - 2.76 \times 10^6 \text{ m}^3)\left(\frac{1 \text{ h}}{3600 \text{ s}}\right)}{1.2 \frac{\text{m}^3}{\text{s}} + 2.8 \frac{\text{m}^3}{\text{s}} - 0.70 \frac{\text{m}^3}{\text{s}} - 2.9 \frac{\text{m}^3}{\text{s}}}$$
$$= 97 \text{ h}$$

$$\Delta t_3 = \frac{(2)(2.81 \times 10^6 \text{ m}^3 - 2.83 \times 10^6 \text{ m}^3)\left(\frac{1 \text{ h}}{3600 \text{ s}}\right)}{2.8 \frac{\text{m}^3}{\text{s}} + 2.5 \frac{\text{m}^3}{\text{s}} - 2.9 \frac{\text{m}^3}{\text{s}} - 3.8 \frac{\text{m}^3}{\text{s}}}$$
$$= 8 \text{ h}$$

t total time to maximum water elevation   h

$$t = \Delta t_1 + \Delta t_2 + \Delta t_3$$
$$= 22 \text{ h} + 97 \text{ h} + 8 \text{ h}$$
$$= 127 \text{ h} \quad (130 \text{ h})$$

The answer is (D).

Why Other Options Are Wrong

(A) This incorrect solution fails to multiply the numerator of the time equation by two and adds instead of subtracts outflow. Other assumptions, definitions, and equations are unchanged from the correct solution.

(B) This incorrect solution adds instead of subtracts the outflow terms in the time increment calculations. Other assumptions, definitions, and equations are unchanged from the correct solution.

(C) This incorrect solution calculates the time based on the initial and final values instead of the incremental data values and reverses the storage terms in the equation. Other assumptions, definitions, and equations are unchanged from the correct solution.

SOLUTION 51

For a paved site where rainfall intensity data is available, the kinematic wave formula can be used for calculating time of concentration. This is a variation of the Manning kinematic equation and uses the Manning roughness coefficient.

i rainfall intensity   in/hr

From the first illustration, the rainfall intensity for a 45 min duration 10-year storm is 3.5 in/hr and for a 45 min 25-year storm is 5.0 in/hr.

L maximum flow path   ft

From the second illustration, the maximum flow path is calculated by

$$L = \sqrt{(1230 \text{ ft})^2 + (390 \text{ ft})^2} = 1290 \text{ ft}$$

n Manning roughness coefficient   –

For smooth impervious surfaces such as paved asphalt surfaces, the Manning roughness coefficient is 0.035.

S average ground surface slope   ft/ft
t_c time of concentration   min

$$t_c = \frac{0.94(nL)^{0.6}}{i^{0.4} S^{0.3}}$$

$$t_{c,10} = \frac{(0.94)(0.035)^{0.6}(1290 \text{ ft})^{0.6}}{\left(3.5 \frac{\text{in}}{\text{hr}}\right)^{0.4} \left(\frac{1.2}{100}\right)^{0.3}}$$
$$= 21.1 \text{ min}$$

$$t_{c,25} = \frac{(0.94)(0.035)^{0.6}(1290 \text{ ft})^{0.6}}{\left(5.0 \frac{\text{in}}{\text{hr}}\right)^{0.4} \left(\frac{1.2}{100}\right)^{0.3}}$$
$$= 18.31 \text{ min}$$

$$\Delta t_c = t_{c,10} - t_{c,25}$$
$$= 21.1 \text{ min} - 18.31 \text{ min}$$
$$= 2.79 \text{ min} \quad (2.8 \text{ min})$$

The answer is (C).

Why Other Options Are Wrong

(A) This incorrect solution uses the slope as a percent instead of as a fraction. Other assumptions, definitions, and equations are the same as used in the correct solution.

(B) This incorrect solution misread the illustration for obtaining rainfall intensity values for the 10-year and 25-year storms. Other assumptions, definitions, and equations are the same as used in the correct solution.

(D) This incorrect solution takes the difference of the rainfall intensities and then applies the time of concentration equation. Other assumptions, definitions, and equations are the same as used in the correct solution.

SOLUTION 52

The average annual demand is 90,000 ac-ft/yr. This is the slope of the average flow line plotted on the following illustration.

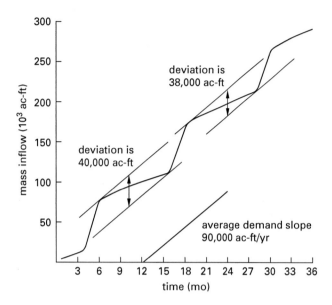

Lines parallel to the average annual demand slope are drawn tangent to the maximum points of deviation for each fill-and-empty cycle of the reservoir. The parallel line pair with the maximum deviation defines the reservoir capacity.

V required capacity ac-ft

$$V = 40{,}000 \text{ ac-ft}$$

The answer is (A).

Why Other Options Are Wrong

(B) This incorrect solution adds the volume defined by the deviation for each fill-and-empty cycle. The illustration is unchanged from the correct solution.

(C) This incorrect solution assumes that the average annual demand and the reservoir capacity are equal.

(D) This incorrect solution assumes that the average annual demand multiplied by the number of years of record is the reservoir capacity.

GROUNDWATER AND WELL FIELDS

SOLUTION 53

g	gravitational constant	9.81 m/s²
k_i	intrinsic permeability	cm²
K_w	hydraulic conductivity for water	cm/s
μ_w	water viscosity	0.01307 g/cm·s at 10°C
ρ_w	water density	0.9997 g/cm³ at 10°C

$$k_i = \frac{K_w \mu_w}{\rho_w g}$$

$$= \frac{\left(2.0 \times 10^{-4} \frac{\text{cm}}{\text{s}}\right)\left(0.01307 \frac{\text{g}}{\text{cm·s}}\right)}{\left(0.9997 \frac{\text{g}}{\text{cm}^3}\right)\left(9.81 \frac{\text{m}}{\text{s}^2}\right)\left(100 \frac{\text{cm}}{\text{m}}\right)}$$

$$= 2.67 \times 10^{-9} \text{ cm}^2$$

K_{NAPL}	hydraulic conductivity of NAPL	cm/s
μ_f	NAPL viscosity	0.066 g/cm·s at 10°C
ρ_f	NAPL density	0.92 g/cm³ at 10°C

$$K_{\text{NAPL}} = \frac{k_i g \rho_f}{\mu_f}$$

$$= \frac{(2.67 \times 10^{-9} \text{ cm}^2)\left(9.81 \frac{\text{m}}{\text{s}^2}\right) \times \left(0.92 \frac{\text{g}}{\text{cm}^3}\right)\left(100 \frac{\text{cm}}{\text{m}}\right)}{0.066 \frac{\text{g}}{\text{cm·s}}}$$

$$= 3.6 \times 10^{-5} \text{ cm/s}$$

The answer is (B).

Why Other Options Are Wrong

(A) This incorrect solution inverts NAPL viscosity and density. Other assumptions, definitions, and equations are unchanged from the correct solution.

(C) This incorrect solution uses density of water instead of NAPL. Other assumptions, definitions, and equations are unchanged from the correct solution.

(D) This incorrect solution uses dynamic viscosity of water instead of NAPL. Other assumptions, definitions, and equations are unchanged from the correct solution.

SOLUTION 54

The groundwater elevation contour lines are drawn to make it possible to determine the groundwater gradient.

i	groundwater gradient	–
ΔL	distance between groundwater contour lines of interest	ft
Δh	elevation change over distance L	ft

$$i = \frac{\Delta h}{\Delta L} = \frac{3210 \text{ ft} - 3207 \text{ ft}}{2300 \text{ ft}} = 0.0013$$

K	hydraulic conductivity	ft/day
n_e	effective porosity	–
r_f	retardation factor	–
v_s	solute velocity	ft/day

$$v_s = \frac{Ki}{n_e r_f} = \frac{\left(0.83\ \dfrac{\text{ft}}{\text{day}}\right)(0.0013)}{(0.37)(1.94)} = 0.0015 \text{ ft/day}$$

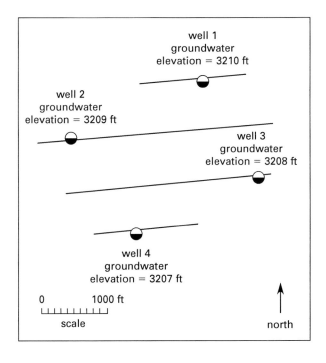

The answer is (B).

Why Other Options Are Wrong

(A) This incorrect solution does not include porosity. This calculates the Darcy velocity, not actual velocity. The figure and other assumptions, definitions, and equations are unchanged from the correct solution.

(C) This incorrect solution multiplies by, instead of divides by, the retardation factor and does not include porosity. The figure and other assumptions, definitions, and equations are unchanged from the correct solution.

(D) This incorrect solution does not include the retardation factor. The figure and other assumptions, definitions, and equations are unchanged from the correct solution.

SOLUTION 55

The dynamic viscosity and density for water at 10°C are 0.001 307 kg/m·s and 999.7 kg/m³, respectively.

g	gravitational constant	m/s²
i	groundwater gradient	–
k	intrinsic permeability	m²
n_e	soil effective porosity	–
v_x	groundwater velocity	m/d
μ_w	water dynamic viscosity	kg/m·s
ρ_w	water density	kg/m³

$$v_x = \frac{kg\rho_w i}{n_e \mu_w}$$

$$= \frac{\left(1.1 \times 10^{-5} \text{mm}^2\right)\left(9.81\ \dfrac{\text{m}}{\text{s}^2}\right)}{(0.38)\left(0.001\,307\ \dfrac{\text{kg}}{\text{m·s}}\right)} \times \left(999.7\ \dfrac{\text{kg}}{\text{m}^3}\right)(0.000\,63)$$

$$\times \left(1000\ \dfrac{\text{mm}}{\text{m}}\right)^2 \left(\dfrac{1\ \text{d}}{86\,400\ \text{s}}\right)$$

$$= 0.0118 \text{ m/d}$$

f_{oc}	organic carbon fraction	–
K_d	distribution coefficient	mL/g
K_{oc}	soil-water partition coefficient	mL/g

$$K_d = K_{oc}f_{oc} = \left(173\ \dfrac{\text{mL}}{\text{g}}\right)\left(485\ \dfrac{\text{mg}}{\text{kg}}\right)\left(\dfrac{1\ \text{kg}}{10^6\ \text{mg}}\right)$$

$$= 0.084 \text{ mL/g}$$

B_d soil bulk density g/cm³
r_f retardation factor –

$$r_f = 1 + \frac{B_d K_d}{n_e}$$

$$= 1 + \frac{\left(1.8 \frac{\text{g}}{\text{cm}^3}\right)\left(0.084 \frac{\text{mL}}{\text{g}}\right)\left(1 \frac{\text{cm}^3}{\text{mL}}\right)}{0.38}$$

$$= 1.4$$

v_s dissolved organic solvent velocity m/d

$$v_s = \frac{v_x}{r_f} = \frac{0.0118 \frac{\text{m}}{\text{d}}}{1.4}$$

$$= 0.0084 \text{ m/d}$$

The answer is (C).

Why Other Options Are Wrong

(A) This incorrect solution uses the intrinsic permeability as the hydraulic conductivity and the water-soil partition coefficient as the distribution coefficient. Other definitions and equations are unchanged from the correct solution.

(B) This incorrect solution uses the Darcy velocity instead of the actual velocity for the groundwater and uses the partition coefficient as the distribution coefficient. Other definitions and equations are unchanged from the correct solution.

(D) This incorrect solution uses the groundwater velocity for the dissolved organic solvent velocity. Other definitions and equations are unchanged from the correct solution.

SOLUTION 56

Assume the release is longitudinally one dimensional, continuous, and at steady state, and assume that diffusion is negligible.

L distance between points of interest m
α_L dispersion coefficient m

$$\alpha_L = 0.0175 L^{1.46} = (0.0175)(1600 \text{ m})^{1.46} = 834 \text{ m}$$

D_L longitudinal hydrodynamic dispersion m²/d
v_x bulk groundwater velocity m/d

$$D_L = \alpha_L v_x = (834 \text{ m})\left(172 \frac{\text{cm}}{\text{d}}\right)\left(\frac{1 \text{ m}}{100 \text{ cm}}\right)$$

$$= 1434 \text{ m}^2/\text{d}$$

C concentration of the solute at time t mg/L
C_o concentration of the solute at time zero mg/L
D_L longitudinal hydrodynamic dispersion m²/d
erfc complementary error function –
t travel time of interest day

$$\frac{2C}{C_o} = \text{erfc}\frac{L - v_x t}{2\sqrt{D_L t}} = \frac{(2)\left(0.001 \frac{\text{mg}}{\text{L}}\right)}{0.182 \frac{\text{mg}}{\text{L}}}$$

$$= 0.01099$$

$$\frac{L - v_x t}{2\sqrt{D_L t}} = 1.66 = \frac{1600 \text{ m} - \left(172 \frac{\text{cm}}{\text{d}}\right)\left(\frac{1 \text{ m}}{100 \text{ cm}}\right)t}{2\sqrt{\left(1434 \frac{\text{m}^2}{\text{d}}\right)t}}$$

Multiply through and square and combine terms, and solve for t using the quadratic formula.

$$t^2 - 7241t + 8.7 \times 10^5 = 0$$

$$t = 122 \text{ d} \quad (120 \text{ d})$$

The answer is (B).

Why Other Options Are Wrong

(A) This incorrect solution fails to convert the groundwater velocity units from cm/d to m/d. Other assumptions, definitions, and equations are the same as used in the correct solution.

(C) This incorrect solution misuses the complementary error function. Other assumptions, definitions, and equations are the same as used in the correct solution.

(D) This incorrect solution misapplies the quadratic formula. Other assumptions, definitions, and equations are the same as used in the correct solution.

SOLUTION 57

d_i layer thickness cm
K overall hydraulic conductivity cm/s
K_i layer hydraulic conductivity cm/s

$$K = \frac{\sum d_i}{\sum \frac{d_i}{K_i}}$$

$$= \frac{130 \text{ cm} + 180 \text{ cm} + 270 \text{ cm} + 65 \text{ cm}}{\frac{130 \text{ cm}}{0.0090 \frac{\text{cm}}{\text{s}}} + \frac{180 \text{ cm}}{0.017 \frac{\text{cm}}{\text{s}}} + \frac{270 \text{ cm}}{0.036 \frac{\text{cm}}{\text{s}}} + \frac{65 \text{ cm}}{0.011 \frac{\text{cm}}{\text{s}}}}$$

$$= 0.0168 \text{ cm/s}$$

A	unit area of aquifer	1 m²
i	hydraulic gradient	m/m, taken as $\Delta h/1$ m
q	vertical ground water flow rate per unit area of aquifer	m³/s·m²

$$q = KiA$$
$$= \left(0.0168 \; \frac{\text{cm}}{\text{s}}\right)\left(14 \; \frac{\text{cm}}{\text{m}}\right)\left(\frac{1 \text{ m}^2}{10\,000 \text{ cm}^2}\right)(1 \text{ m}^2)$$
$$= 2.35 \times 10^{-5} \text{ m}^3/\text{s for 1 m}^2 \text{ of aquifer area}$$

The answer is (B).

Why Other Options Are Wrong

(A) This incorrect solution uses the average of the hydraulic conductivities and the ratio of the thicknesses of the five layers to that of the four layers. Other assumptions, definitions, and equations are unchanged from the correct solution.

(C) This incorrect solution includes all layers in the calculation for overall hydraulic conductivity, not only those layers defining the flow path between the screened intervals of the two wells. Other assumptions, definitions, and equations are unchanged from the correct solution.

(D) This incorrect solution uses the average hydraulic conductivity of the soil layers. Other assumptions, definitions, and equations are unchanged from the correct solution.

SOLUTION 58

d_w	well casing diameter	ft
h, h_o	height of drawdown curve measured from the top of confined aquifer at radial distance $r + r_w$ and $r_o + r_w$ from the well	ft
K	hydraulic conductivity	ft/day
m	thickness of the confined aquifer	ft
Q	discharge flow rate	ft³/sec
r, r_o	radial distances from the well casing	ft
r_w	well casing radius	ft

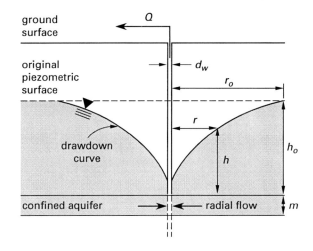

$$r_w = \frac{d_w}{2} = \frac{1 \text{ ft}}{2} = 0.5 \text{ ft}$$

$$Q = \frac{2\pi K m(h_o - h)}{\ln \frac{r_o + r_w}{r + r_w}}$$

$$= \frac{2\pi \left(7.3 \; \frac{\text{ft}}{\text{day}}\right)(20 \text{ ft})(105 \text{ ft} - 70 \text{ ft}) \times \left(\frac{1 \text{ day}}{86{,}400 \text{ sec}}\right)}{\ln\left(\frac{120 \text{ ft} + 0.5 \text{ ft}}{50 \text{ ft} + 0.5 \text{ ft}}\right)}$$

$$= 0.43 \text{ ft}^3/\text{sec}$$

The answer is (C).

Why Other Options Are Wrong

(A) This incorrect solution improperly applies the discharge flow equation. The illustration and other definitions and equations used in the correct solution are unchanged.

(B) This incorrect solution neglects to include the natural log function in the discharge equation. The illustration and other definitions and equations used in the correct solution are unchanged.

(D) This incorrect solution uses the discharge flow equation for an unconfined aquifer instead of the equation for a confined aquifer. Other definitions used in the correct solution are unchanged.

SOLUTION 59

d	distance between bottom of drains and top of impermeable layer	m
h	water depth in the drain	m
H	vertical distance between the bottom of the drain and the highest point of the piezometric surface	m
K	hydraulic conductivity	cm/s
q	infiltration rate	cm/s
S	drain spacing	m

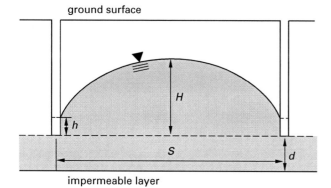

$$S^2 = \frac{4K(H^2 - h^2 + 2dH - 2dh)}{q}$$

$$= \frac{(4)\left(0.018 \ \frac{\text{cm}}{\text{s}}\right)\left((1.8 \text{ m})^2 - (0.25 \text{ m})^2 \right. }{0.0014 \ \frac{\text{cm}}{\text{s}}}$$

$$\frac{\left. + (2)(0.4 \text{ m})(1.8 \text{ m}) - (2)(0.4 \text{ m})(0.25 \text{ m})\right)}{}$$

$$= 227 \text{ m}^2$$

$$S = \sqrt{227 \text{ m}^2}$$

$$= 15 \text{ m}$$

The answer is (B).

Why Other Options Are Wrong

(A) This incorrect solution reverses the values for the hydraulic conductivity (K) and infiltration rate (q) in the spacing equation. The illustration and other definitions are unchanged from the correct solution.

(C) This incorrect solution fails to square the drain spacing (S), vertical distance (H) and water depth (h) terms. The illustration and other definitions are unchanged from the correct solution.

(D) This incorrect solution fails to take the square root of the S^2 term. The illustration and other definitions are unchanged from the correct solution.

WASTEWATER TREATMENT

SOLUTION 60

To find the recirculated solids flow rate, perform a mass balance around the clarifier based on biomass. A mass balance based on substrate will produce a trivial solution.

Q	flow rate	10^6 gal/day
Q_r	recirculated solids flow rate	10^6 gal/day
Q_w	wasted solids flow rate	10^6 gal/day
X	reactor mixed liquor suspended solids concentration	mg/L
X_e	clarifier effluent solids concentration	mg/L
X_u	recirculated solids concentration	mg/L

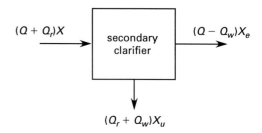

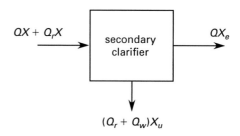

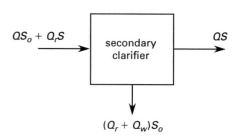

The mass of biomass equals the mass of biomass recirculated plus the mass of biomass wasted.

$$(Q + Q_r)X = (Q_r + Q_w)X_u + (Q - Q_w)X_e$$

Assume the clarifier effluent solids concentration is negligible compared to the reactor mixed liquor and recirculated solids concentrations.

$$X_e \ll X, X_u$$
$$(Q - Q_w)X_e \approx 0$$
$$(Q + Q_r)X = (Q_r + Q_w)X_u$$

Multiply through and reorder the equation in terms of the recirculated solids flow rate, Q_r.

$$QX + Q_rX = Q_rX_u + Q_wX_u$$
$$Q_r(X - X_u) = Q_wX_u - QX$$
$$Q_r = \frac{Q_wX_u - QX}{X - X_u}$$
$$= \frac{QX - Q_wX_u}{X_u - X}$$

All parameters are known except the wasted solids and recirculated solids flow rates. The wasted solids flow rate can be calculated from the definition of mean cell residence time.

V	reactor volume	10^6 gal
θ_c	mean cell residence time	day

$$\theta_c = \frac{VX}{Q_wX_u}$$
$$Q_w = \frac{VX}{\theta_cX_u}$$
$$= \frac{(5 \times 10^6 \text{ gal})\left(3500 \frac{\text{mg}}{\text{L}}\right)}{(10 \text{ day})\left(12\,000 \frac{\text{mg}}{\text{L}}\right)}$$
$$= 1.46 \times 10^5 \text{ gal/day}$$

$$Q_r = \frac{\left(2.5 \times 10^7 \frac{\text{gal}}{\text{day}}\right)\left(3500 \frac{\text{mg}}{\text{L}}\right) - \left(1.46 \times 10^5 \frac{\text{gal}}{\text{day}}\right)\left(12\,000 \frac{\text{mg}}{\text{L}}\right)}{12\,000 \frac{\text{mg}}{\text{L}} - 3500 \frac{\text{mg}}{\text{L}}}$$
$$= 1.0 \times 10^7 \text{ gal/day}$$

The answer is (A).

Why Other Options Are Wrong

(B) This incorrect solution is based on an improperly defined mass balance. The mass balance is performed around the clarifier using biomass, but the labeling of biomass inputs and outputs is incorrect. Other assumptions and definitions are unchanged from the correct solution.

(C) This incorrect solution is based on an improperly defined mass balance. The mass balance is performed around the clarifier, but uses substrate instead of biomass and the substrate labeling is wrong. Other definitions and assumptions are unchanged from the correct solution.

(D) This incorrect solution applies the "plug-flow" equation for activated sludge solved for recirculation ratio. Other assumptions and definitions are unchanged from the correct solution.

SOLUTION 61

f	solids fraction	–
$\dot{m}$	solids dry mass flow rate	lbm/day
$\dot{V}$	volumetric flow rate	gal/day
ρ	wet sludge density	lbm/ft^3

$$\dot{m} = f_1\rho\dot{V}_1 = f_2\rho\dot{V}_2$$
$$\dot{V}_2 = \frac{f_1\dot{V}_1}{f_2}$$

At 1.2% solids,
$$\dot{V}_1 = 50{,}000 \text{ gal/day}$$

At 24% solids,
$$\dot{V}_2 = \frac{(0.012)\left(50{,}000 \frac{\text{gal}}{\text{day}}\right)}{0.24}$$
$$= 2500 \text{ gal/day}$$

$\dot{V}_r$ volume flow rate reduction realized gal/day

$$\dot{V}_r = \dot{V}_1 - \dot{V}_2$$
$$= 50{,}000 \frac{\text{gal}}{\text{day}} - 2500 \frac{\text{gal}}{\text{day}}$$
$$= 47{,}500 \text{ gal/day} \quad (48{,}000 \text{ gal/day})$$

The answer is (D).

Why Other Options Are Wrong

(A) This incorrect choice is the calculated sludge volume flow rate at 24% solids. The difference between the volume flow rate at 24% solids and 1.2% solids is required for the correct answer. Other equations and definitions are the same as used in the correct solution.

(B) This incorrect choice uses the difference between 24% and 1.2% solids and solids mass, and fails to use the mass/density/volume relationship consistently. Other equations and definitions are the same as used in the correct solution.

(C) This incorrect choice does not conserve mass. Other equations and definitions are the same as used in the correct solution.

SOLUTION 62

K_H	Henry's constant	atm
K'_H	Henry's constant	–
MW	molecular weight	g/mol
R^*	universal gas constant	0.082 atm·L/mol·K
T	temperature	K
ρ_w	water density	1000 g/L

The Henry's constant for methylene chloride is 177 atm.

$$K'_H = \frac{K_H \text{MW}_w}{\rho_{\text{water}} R^* T}$$

$$= \frac{(177 \text{ atm})\left(18 \frac{\text{g}}{\text{mol}}\right)}{\left(1000 \frac{\text{g}}{\text{L}}\right)\left(0.082 \frac{\text{atm·L}}{\text{mol·K}}\right)(25°\text{C} + 273°)}$$

$$= 0.13 \text{ unitless}$$

S	stripping factor	–
V_a/V_w	air-to-water ratio	–

$$\frac{V_a}{V_w} = \frac{S}{K'_H} = \frac{3.5}{0.13}$$

$$= 27$$

Q	water flow rate	gal/min
Q_a	air flow rate	ft³/min

$$Q_a = Q\left(\frac{V_a}{V_w}\right)$$

$$= \left(135 \frac{\text{gal}}{\text{min}}\right)\left(0.134 \frac{\text{ft}^3}{\text{gal}}\right)(27)$$

$$= 488 \text{ ft}^3/\text{min} \quad (490 \text{ ft}^3/\text{min})$$

The answer is (C).

Why Other Options Are Wrong

(A) This incorrect choice includes the concentration ratio in the calculation of air-to-water ratio. Other assumptions, definitions, and equations are unchanged from the correct solution.

(B) This incorrect choice uses the stripping factor as the air-to-water ratio. Other definitions are unchanged from the correct solution.

(D) This incorrect choice uses Henry's constant with units of atm instead of unitless and inverts the air-to-water ratio. Other assumptions, definitions, and equations are the same as used for the correct solution.

SOLUTION 63

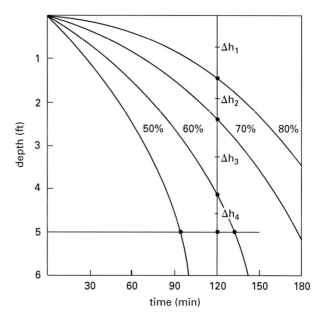

D	total depth	ft
E	total efficiency	%
R_i	incremental efficiency	%
Δh_i	incremental change in depth	ft

$$E = \sum 0.5 \left(\frac{\Delta h_i}{d}\right)(R_i + R_{i+1})$$

$$= (0.5)\left(\frac{1.42 \text{ ft}}{5 \text{ ft}}\right)(100\% + 80\%)$$

$$+ (0.5)\left(\frac{0.95 \text{ ft}}{5 \text{ ft}}\right)(80\% + 70\%)$$

$$+ (0.5)\left(\frac{1.68 \text{ ft}}{5 \text{ ft}}\right)(70\% + 60\%)$$

$$+ (0.5)\left(\frac{0.82 \text{ ft}}{5 \text{ ft}}\right)(60\% + 57\%)$$

$$= 25.56\% + 14.25\% + 21.84\% + 9.59\%$$

$$= 71\%$$

The answer is (D).

Why Other Options Are Wrong

(A) This incorrect solution uses the efficiency at the height-time coordinate of 5 ft and 120 min subtracted from 100%.

(B) This incorrect solution uses the efficiency as the height-time coordinate for 5 ft and 120 min.

(C) This incorrect solution does not average the incremental efficiencies. The illustration and definitions are unchanged from the correct solution.

SOLUTION 64

The winter loading rate will control because it results in the largest surface area.

C	BOD concentration	mg/L
$\dot{V}$	volumetric flow rate	gal/day
$\dot{m}$	mass flow rate	lbm/day

$$\dot{m} = \dot{V}C$$
$$= \left(2.6 \times 10^5 \frac{\text{gal}}{\text{day}}\right)\left(14\,000 \frac{\text{mg BOD}}{\text{L}}\right)$$
$$\times \left(3.785 \frac{\text{L}}{\text{gal}}\right)\left(2.204 \frac{\text{lbm}}{10^6 \text{ mg}}\right)$$
$$= 30{,}365 \text{ lbm BOD/day}$$

A	surface area	ac
D	liquid depth	ft
OLR	organic loading rate	lbm/10^3 ft^3-day

$$A = \frac{\dot{m}}{(\text{OLR})D}$$
$$= \frac{\left(30{,}365 \frac{\text{lbm BOD}}{\text{day}}\right)\left(\frac{1 \text{ ac}}{43{,}560 \text{ ft}^2}\right)}{\left(12 \frac{\text{lbm BOD}}{10^3 \text{ ft}^3\text{-day}}\right)(10 \text{ ft})}$$
$$= 5.8 \text{ ac}$$

The answer is (D).

Why Other Options Are Wrong

(A) This incorrect solution calculates the surface area based on summer loading rate. Other definitions and equations are unchanged from the correct solution.

(B) This incorrect solution calculates the surface area based on the average of the winter and summer loading rates.

(C) This incorrect solution includes an error in the conversion factor from liters to gallons. Other definitions and equations are unchanged from the correct solution.

SOLUTION 65

k_d	endogenous decay rate constant	d^{-1}
Y	yield coefficient	mg/mg
Y_{obs}	observed yield coefficient	mg/mg
θ_c	mean cell residence time	d

$$Y_{\text{obs}} = \frac{Y}{1 + k_d \theta_c} = \frac{0.53 \frac{\text{g}}{\text{g}}}{1 + \left(\frac{0.05}{\text{d}}\right)(8 \text{ d})}$$
$$= 0.38 \text{ g/g}$$

Q	influent flow rate	m^3/d
S	effluent BOD	mg/L
S_o	influent BOD	mg/L
X_p	mass of biomass produced	kg/d

$$X_p = Y_{\text{obs}}(S_o - S)Q$$
$$= \left(0.38 \frac{\text{g}}{\text{g}}\right)\left(281 \frac{\text{mg}}{\text{L}} - 20 \frac{\text{mg}}{\text{L}}\right)\left(27\,000 \frac{\text{m}^3}{\text{d}}\right)$$
$$\times \left(1000 \frac{\text{L}}{\text{m}^3}\right)\left(10^{-6} \frac{\text{kg}}{\text{mg}}\right)$$
$$= 2678 \text{ kg/d} \quad (2700 \text{ kg/d})$$

The answer is (A).

Why Other Options Are Wrong

(B) This incorrect choice does not correct the yield coefficient for cell death. Other equations are unchanged from the correct solution.

(C) This incorrect choice does not include the yield coefficient in the calculation. Other definitions are unchanged from the correct solution.

(D) This incorrect choice miscalculates the observed yield coefficient. Other definitions and equations are unchanged from the correct solution.

SOLUTION 66

Check the surface area required based on hydraulic loading. Typical hydraulic loading rates vary between 2.0 and 4.0 gal/ft^2-day. To be conservative, use the minimum value of 2.0 gal/ft^2-day.

A_h	media surface area based on hydraulic loading	ft^2
HLR	hydraulic loading rate	gal/ft^2-day
Q	influent flow rate	gal/day

$$A_h = \frac{Q}{\text{HLR}} = \frac{250{,}000 \frac{\text{gal}}{\text{day}}}{2.0 \frac{\text{gal}}{\text{ft}^2\text{-day}}}$$
$$= 125{,}000 \text{ ft}^2$$

Check the surface area required based on organic loading. Typical organic loading rates vary between 2.0 and 3.5 lbm total BOD/10^3 ft^2-day. To be conservative, use the minimum value of 2.0 lbm/10^3 ft^2-day.

A_s	media surface area based on organic loading	ft^2
OLR	organic loading rate	lbm/10^3 ft^2-day
S_o	influent total BOD	mg/L

$$A_s = \frac{QS_o}{\text{OLR}}$$

$$= \frac{\left(250{,}000\ \dfrac{\text{gal}}{\text{day}}\right)\left(174\ \dfrac{\text{mg}}{\text{L}}\right) \times \left(2.204\ \dfrac{\text{lbm}}{10^6\ \text{mg}}\right)\left(3.785\ \dfrac{\text{L}}{\text{gal}}\right)}{2.0\ \dfrac{\text{lbm}}{10^3\ \text{ft}^2\text{-day}}}$$

$$= 181{,}442\ \text{ft}^2\quad (180{,}000\ \text{ft}^2)$$

$$A_s > A_h$$

Total media surface area is 180,000 ft².

The answer is (C).

Why Other Options Are Wrong

(A) This incorrect option does not check surface area based on organic loading rate. Other assumptions, definitions, and equations are the same as those used in the correct solution.

(B) This incorrect option subtracts the effluent BOD from the influent BOD in the calculation for surface area based on organic loading. Other assumptions, definitions, and equations are the same as those used in the correct solution.

(D) This incorrect option includes the wrong conversion factor for consistent volume units. Other assumptions, definitions, and equations are the same as used in the correct solution.

SOLUTION 67

An SF wetland system may be selected over an FWS wetland system for reasons of improved odor and vector control, reduced public exposure, improved suspended solids removal, lower susceptibility to climatic temperature extremes, and greater reaction rates. However, because the exposed water surface provides a better opportunity for oxygen transfer, the FWS system may be preferred where an aerobic environment is necessary to effect a desired treatment such as nitrification for ammonia removal.

The answer is (C).

Why Other Options Are Wrong

(A) This choice is incorrect because the SF wetland is operated without exposing the water surface. The submerged flow limits vector access and odor dispersion by wind currents.

(B) This choice is incorrect because the SF wetland provides an opportunity for filtration as the water passes through the media and, because the water is not exposed to sunlight, reduces the occurrence of algae in the effluent. Both of these contribute to improved suspended solids removal.

(D) This choice is incorrect because by not exposing the water surface to the air, the SF wetland experiences less impact from ambient air temperature fluctuations and other climatic variations.

SOLUTION 68

For a complete mix reactor without recycle, the hydraulic residence time and the mean cell residence time are equal.

k_d	endogenous decay rate coefficient	d^{-1}
S	effluent biochemical oxygen demand (BOD)	mg/L
S_o	influent BOD	mg/L
X	mixed liquor suspended solids (MLSS)	mg/L
Y	yield coefficient	g/g
θ	hydraulic residence time	d
θ_c	mean cell residence time	d

Values for the yield coefficient and the endogenous decay-rate coefficient are found from the following linear equation. The yield coefficient is the slope and the endogenous decay rate coefficient is the intercept.

$$\frac{1}{\theta_c} = \frac{Y(S_o - S)}{\theta X} - k_d$$

$1/\theta_c$ (d^{-1})	$(S_o - S)/\theta X$ (d^{-1})	equation
0.29	0.49	$0.29 = Y\,0.49 - k_d$
0.52	0.75	$0.52 = Y\,0.75 - k_d$
0.69	0.95	$0.69 = Y\,0.95 - k_d$
0.91	1.2	$0.91 = Y\,1.2 - k_d$

Solving any two of the equations simultaneously will give values for the yield coefficient and the endogenous decay-rate coefficient.

$$\frac{0.29}{\text{d}} = Y\frac{0.49}{\text{d}} - k_d$$

$$Y = \frac{\dfrac{0.29}{\text{d}} + k_d}{\dfrac{0.49}{\text{d}}}$$

$$\frac{0.69}{\text{d}} = \frac{\left(\dfrac{0.29}{\text{d}} + k_d\right)\left(\dfrac{0.95}{\text{d}}\right)}{\dfrac{0.49}{\text{d}}} - k_d$$

$$= \frac{\left(\frac{0.95}{d}\right)\left(\frac{0.29}{d}\right)}{\frac{0.49}{d}} + \frac{\frac{0.95}{d}k_d}{\frac{0.49}{d}} - k_d$$

$$= \frac{0.56}{d} + \frac{1.9}{d}k_d - k_d$$

$$k_d = \frac{\frac{0.69}{d} - \frac{0.56}{d}}{\frac{0.9}{d}}$$

$$= 0.14 \text{ d}^{-1}$$

The answer is (C).

Why Other Options Are Wrong

(A) This incorrect solution makes a mathematical error when solving the two equations. Other assumptions, definitions, and equations are unchanged from the correct solution.

(B) This incorrect solution assumes a mean cell residence time of 10 d. Other definitions and equations are unchanged from the correct solution.

(D) This incorrect solution solves for the yield coefficient. Other assumptions, definitions, and equations are unchanged from the correct solution.

SOLUTION 69

f	ratio of BOD$_5$ to BOD$_u$	
N	effluent ammonia concentration	mg/L
N_o	influent ammonia concentration	mg/L
S	effluent BOD$_5$	mg/L
S_o	influent BOD$_5$	mg/L
Q	flow rate	gal/day
X_p	biomass wasted	lbm/day

The daily oxygen requirement must include both oxidation of the BOD and nitrification.

$$\frac{\text{lbm O}_2}{\text{day}} = Q(S_o - S)f - 1.42X_p + 4.57Q(N_o - N)$$

The 1.42 accounts for the theoretical oxygen demand for mineralizing the wasted biomass and the 4.57 accounts for the nitrogenous oxygen demand (4.57 g oxygen per gram of ammonia).

$$\frac{\text{lbm O}_2}{\text{day}} = \left(5.0 \times 10^6 \frac{\text{gal}}{\text{day}}\right)\left(3.785 \frac{\text{L}}{\text{gal}}\right)$$
$$\times \left(356 \frac{\text{mg}}{\text{L}} - 30 \frac{\text{mg}}{\text{L}}\right)(1.52)\left(\frac{2.204 \text{ lbm}}{10^6 \text{ mg}}\right)$$
$$- (1.42)\left(2800 \frac{\text{lbm}}{\text{day}}\right) + (4.57)$$
$$\times \left(5.0 \times 10^6 \frac{\text{gal}}{\text{day}}\right)\left(3.785 \frac{\text{L}}{\text{gal}}\right)$$
$$\times \left(\frac{2.204 \text{ lbm}}{10^6 \text{ mg}}\right)\left(63 \frac{\text{mg}}{\text{L}} - 10 \frac{\text{mg}}{\text{L}}\right)$$
$$= 26{,}795 \text{ lbm/day} \quad (27{,}000 \text{ lbm/day})$$

The answer is (C).

Why Other Options Are Wrong

(A) This incorrect solution fails to include the nitrogenous oxygen demand. Other assumptions, definitions, and equations are the same as used in the correct solution.

(B) This incorrect solution fails to convert BOD$_5$ to BOD$_u$. Other assumptions, definitions, and equations are the same as used in the correct solution.

(D) This incorrect solution fails to subtract the theoretical oxygen demand for mineralizing biomass. Other assumptions, definitions, and equations are the same as used in the correct solution.

SOLUTION 70

k	growth rate constant	d^{-1}
Y	yield coefficient	g/g
μ_m	maximum growth rate	d^{-1}

$$\mu_m = kY = \left(\frac{2.1}{\text{d}}\right)\left(0.23 \frac{\text{g}}{\text{g}}\right)$$
$$= 0.48 \text{ d}^{-1}$$

DO	dissolved oxygen concentration	mg/L
K_o	half-saturation constant for DO	mg/L
T	temperature	°C
μ'_m	corrected maximum growth rate	d^{-1}

Assume that a typical value for K_o is 1.3 mg/L.

$$\mu'_m = \mu_m e^{(0.098)(T-15)}\left(\frac{\text{DO}}{K_o + \text{DO}}\right)$$
$$\times (1 - 0.833(7.2 - \text{pH}))$$

$$= \left(\frac{0.48}{\text{d}}\right)(e^{(0.098)(17-15)})\left(\frac{7.2 \frac{\text{mg}}{\text{L}}}{1.3 \frac{\text{mg}}{\text{L}} + 7.2 \frac{\text{mg}}{\text{L}}}\right)$$
$$\times (1 - (0.833)(7.2 - 6.4))$$
$$= 0.17 \text{ d}^{-1}$$

The answer is (B).

Why Other Options Are Wrong

(A) This incorrect solution assumes a minimum dissolved oxygen concentration of 2.0 mg/L and uses it in place of the given value. Other assumptions, definitions, and equations are unchanged from the correct solution.

(C) This incorrect solution calculates the maximum growth, but does not apply corrections. Other assumptions, definitions, and equations are unchanged from the correct solution.

(D) This incorrect solution uses the value of the growth rate constant for the maximum growth rate. Other assumptions, definitions, and equations are unchanged from the correct solution.

SOLUTION 71

Using the chemical formula for biomass, for every 5 mol of influent carbon, 1 mol of nitrogen is required. The chemical formula, molecular weight, and concentration of acetic acid determine the moles of influent carbon.

The molecular formula for acetic acid is CH_3COOH.

The mole weight of acetic acid is

$$(2)\left(12 \frac{g}{mol}\right) + (4)\left(1 \frac{g}{mol}\right) + (2)\left(16 \frac{g}{mol}\right)$$
$$= 60 \text{ g/mol}$$

The ratio of carbon to acetic acid in the influent is 2 mol C/1 mol acetic acid. The daily molar quantity of carbon in the wastewater is

$$\frac{\left(\frac{2}{1}\right)\left(310 \frac{mg}{L}\right)\left(\frac{1 \text{ g}}{1000 \text{ mg}}\right)\left(12\,000 \frac{m^3}{d}\right)}{\left(60 \frac{g}{mol}\right)\left(\frac{1 \text{ m}^3}{1000 \text{ L}}\right)}$$
$$= 124\,000 \text{ mol/d}$$

The ratio of nitrogen to carbon in the biomass chemical formula is used to find the moles of nitrogen required.

$$\frac{1 \text{ mol N}}{5 \text{ mol C}} = \frac{\frac{\text{mol N}}{d}}{124\,000 \frac{\text{mol C}}{d}}$$

$$\frac{\text{mol N}}{d} = \frac{\left(124\,000 \frac{\text{mol C}}{d}\right)(1 \text{ mol N})}{5 \text{ mol C}}$$
$$= 24\,800 \text{ mol/d}$$

Multiplying the moles of nitrogen required per day by the molecular weight of nitrogen gives the daily mass of nitrogen required.

$$\frac{\left(24\,800 \frac{mol}{d}\right)\left(14 \frac{g}{mol}\right)\left(\frac{1 \text{ kg}}{1000 \text{ g}}\right)}{}$$
$$= 347.2 \text{ kg/d} \quad (350 \text{ kg/d})$$

The answer is (C).

Why Other Options Are Wrong

(A) This incorrect solution confuses the molecular weights of carbon and nitrogen.

(C) This incorrect solution uses the wrong chemical formula for acetic acid.

(D) This incorrect solution uses the ratio of the biomass nitrogen to acetic acid carbon instead of the biomass nitrogen to biomass carbon.

SOLUTION 72

C_f adsorber effluent concentration mg/L

$$C_f = (1 - 0.99)\left(148 \frac{mg}{L}\right) = 1.48 \text{ mg/L}$$

C_o initial concentration mg/L
k isotherm intercept mg/g
$1/n$ isotherm slope 0.45
M granular activated carbon (GAC) mass-use rate g/L

$$M = \frac{C_o - C_f}{kC_f^{1/n}} = \frac{148 \frac{mg}{L} - 1.48 \frac{mg}{L}}{\left(220 \frac{mg}{g}\right)\left(1.48 \frac{mg}{L}\right)^{0.45}}$$
$$= 0.56 \text{ g GAC/L water treated}$$

$\dot{m}$ mass flow rate lbm/day
$\dot{V}$ volumetric flow rate gal/day

The daily GAC use rate is

$$\dot{m} = M\dot{V}$$
$$= \left(0.56 \frac{g}{L}\right)\left(2.5 \times 10^5 \frac{gal}{day}\right)\left(3.785 \frac{L}{gal}\right)$$
$$\times \left(\frac{2.204 \text{ lbm}}{1000 \text{ g}}\right)$$
$$= 1168 \text{ lbm/day}$$

Standard sizes of adsorption vessels are 2000 lbm, 4000 lbm, 10,000 lbm, and 20,000 lbm.

The number of days to saturation is

$$\frac{20{,}000 \text{ lbm}}{1168 \frac{lbm}{day}} = 17 \text{ day} > 14 \text{ day}$$

Only the 20,000 lbm vessel will provide a minimum GAC change-out period of 14 days.

The answer is (D).

Why Other Options Are Wrong

(A) This incorrect solution miscalculates the effluent concentration. Other definitions, equations, and standard vessel sizes are the same as for the correct solution.

(B) This incorrect solution confuses the units in the pounds mass to grams conversion factor. Other definitions, equations, and standard vessel sizes are the same as the correct solution.

(C) This incorrect solution uses the inverse slope in the isotherm equation. Other definitions, equations, and standard vessel sizes are the same as the correct solution.

SOLUTION 73

Assume that the neutral solution has a pH of 7.0.

$$\text{pH} + \text{pOH} = 14$$

At pH of 1.6,

$$\text{pOH} = 14 - 1.6 = 12.4$$
$$[\text{OH}^-] = 10^{-12.4} \text{ mol/L} \quad (4.0 \times 10^{-13} \text{ mol/L})$$

At pH of 7.0,

$$\text{pOH} = 14 - 7.0 = 7.0$$
$$[\text{OH}^-] = 10^{-7.0} \text{ mol/L} \quad (1.0 \times 10^{-7} \text{ mol/L})$$
$$\text{OH}^- \text{ required} = 1.0 \times 10^{-7} \, \frac{\text{mol}}{\text{L}} - 4.0 \times 10^{-13}$$
$$= 1.0 \times 10^{-7} \text{ mol/L}$$
$$\text{NaOH} \rightarrow \text{Na}^+ + \text{OH}^-$$

One mole of NaOH dissociates to produce one mole of OH^-.

$$\text{NaOH required} = 1.0 \times 10^{-7} \text{ mol/L}$$

The daily sodium hydroxide feed rate is

$$\frac{\left(50{,}000 \, \frac{\text{gal waste}}{\text{day}}\right)\left(1.0 \times 10^{-7} \, \frac{\text{mol NaOH}}{\text{L waste}}\right)}{0.005 \, \frac{\text{meq}}{\text{L acid}}} \times \left(1 \, \frac{\text{meq}}{\text{mol NaOH}}\right)\left(3785 \, \frac{\text{mL}}{\text{gal}}\right)$$

$$= 3.8 \times 10^3 \text{ mL/d}$$

The answer is (B).

Why Other Options Are Wrong

(A) This incorrect solution bases the hydroxide concentration on the pOH for a corresponding pH of 1.6 and does not consider the pOH under neutralized conditions. Other assumptions are unchanged from the correct solution.

(C) This incorrect solution misuses the definition of pOH to find the hydroxide concentration. Other assumptions are unchanged from the correct solution.

(D) This incorrect solution confuses the relationship between the acid requirement and its corresponding concentration. Other assumptions are unchanged from the correct solution.

SOLUTION 74

To reduce resin deterioration, the influent solution must be diluted to a maximum concentration of 11%.

C_1	undiluted concentration	%
C_2	diluted concentration	%
Q_D	dilution flow rate	m³/d
Q_w	waste flow rate	m³/d
V_D	fractional dilution volume	–

For the feed,

$$V_D = \frac{C_1}{C_2} - 1 = \frac{15\%}{11\%} - 1 = 0.36$$

To meet reuse specifications, the recovered solution must be diluted to 3.5 molar.

The mole weight of CrO_3 is

$$52 \, \frac{\text{g}}{\text{mol}} + (3)\left(16 \, \frac{\text{g}}{\text{mol}}\right) = 100 \text{ g/mol}$$

$$Q_D = V_D Q_w = (0.36)\left(1000 \, \frac{\text{m}^3}{\text{d}}\right)$$
$$= 360 \text{ m}^3/\text{d}$$

For the recovered solution, the equivalent percent concentration of 3.5 molar is

$$\left(3.5 \, \frac{\text{mol}}{\text{L}}\right)\left(100 \, \frac{\text{g}}{\text{mol}}\right)\left(\frac{1 \text{ L}}{1000 \text{ g}}\right) \times 100\% = 35\%$$

$$V_D = \frac{42\%}{35\%} - 1 = 0.20$$

$$Q_D = V_D Q_w = (0.20)\left(350 \, \frac{\text{m}^3}{\text{d}}\right)$$
$$= 70 \text{ m}^3/\text{d}$$

For total dilution,

$$Q_D = 360 \, \frac{\text{m}^3}{\text{d}} + 70 \, \frac{\text{m}^3}{\text{d}} = 430 \text{ m}^3/\text{d}$$

The answer is (A).

(B) This incorrect choice bases the dilution requirement on the ratio of 15% to 3.5 molar. Other assumptions, definitions, and equations are the same as the correct solution.

(C) This incorrect choice defines the fractional dilution volume as the ratio of the undiluted to the diluted concentrations. Other assumptions, definitions, and equations are the same as the correct solution. For the feed,

(D) This incorrect choice bases the dilution requirement on the ratio of 11% to 3.5 molar. Other assumptions, definitions, and equations are the same as the correct solution.

SOLUTION 75

Assume that the contaminant is acetate and that the ammonia concentration is as ammonium (NH_4^+).

The mole weight of NH_4^+ is

$$14 \, \frac{g}{mol} + (4)\left(1 \, \frac{g}{mol}\right) = 18 \, g/mol$$

The mole weight of CH_3COO^- is

$$(2)\left(12 \, \frac{g}{mol}\right) + (3)\left(1 \, \frac{g}{mol}\right) + (2)\left(16 \, \frac{g}{mol}\right)$$
$$= 59 \, g/mol$$

The molar concentration of ammonium is

$$\frac{\left(9.7 \, \frac{mg \, NH_4^+}{L}\right)\left(\frac{1 \, g}{1000 \, mg}\right)}{18 \, \frac{g}{mol}} = 0.00054 \, \frac{mol \, NH_4^+}{L}$$

Six mol of ammonium will react with 26 mol of acetate. Using this ratio, the mass concentration of acetate is

$$\frac{\left(0.00054 \, \frac{mol \, NH_4^+}{L}\right)(26 \, mol \, CH_3COO^-)}{(6 \, mol \, NH_4^+)} \times \left(59 \, \frac{g}{mol}\right)\left(\frac{1000 \, mg}{g}\right)$$
$$= 138 \, mg/L \quad (140 \, mg/L)$$

The answer is (C).

Why Other Options Are Wrong

(A) This incorrect solution inverts the mole ratios of ammonium and acetate. Other assumptions are the same as used in the correct solution.

(B) This incorrect solution calculates the ammonia requirement. Assumptions are the same as used in the correct solution.

(D) This incorrect solution mistakenly uses the biomass as the contaminant and calculates the biomass produced by the chemical biodegradation. Other assumptions are the same as used in the correct solution.

WATER QUALITY

SOLUTION 76

The minimum standards for secondary wastewater treatment included in the CWA place effluent limits on five day biochemical oxygen demand (BOD_5), suspended solids, and hydrogen-ion concentration (pH). In some cases carbonaceous BOD_5 may be substituted for BOD_5. Although disinfection of effluents is required, the specific concerns of disinfection byproducts and dissolved solids are not included in the minimum standards for secondary treatment of wastewater, both being primarily associated with drinking water.

The answer is (A).

Why Other Options Are Wrong

(B) This response is incorrect because, although suspended solids are included, secondary wastewater treatment standards do not include disinfection byproducts.

(C) This response is incorrect because secondary wastewater treatment standards for solids are limited to suspended solids and do not include dissolved solids.

(D) This response is incorrect because neither disinfection byproducts nor dissolved solids are included among the minimum standards for secondary treatment of wastewater.

SOLUTION 77

Bacteria and viruses are small organisms that, unless flocculated or sorbed to larger particles, will pass through a filter bed. These are commonly targeted in application of chlorine for disinfection and are routinely analyzed using indicator organisms and other methods. Protozoa include *Giardia lamblia* and *cryptosporidium* and helminths include pinworms, roundworms, and tapeworms. Protozoan oocysts protect the organisms during chlorination. Although less resistant to the effects of chlorination than protozoa, helminths and helminth ova along with protozoa are primarily removed by filtration. Where filtration is not practiced as part of wastewater treatment, large populations of helminths and protozoa may be discharged to receiving waters.

The answer is (D).

Why Other Options Are Wrong

(A) This option is incorrect because both protozoa and helminths are infrequently looked for in routine analysis, are generally less susceptible to chlorination, and are more likely to be targeted for removal by filtration.

(B) This option is incorrect because both viruses and bacteria are readily susceptible to chlorination, are the object of routine analyses, and, unless sorbed to other particles or flocculated, are not targeted by filtration.

(C) This option is incorrect because viruses are not generally resistant to chlorination or effectively removed by filtration, unless sorbed to larger particles or flocculated. Helminths are not included in routine analyses and are removed by filtration. Also, helminths may be less affected by chlorination than viruses are.

SOLUTION 78

Typically bottles outside the range of 2.0 mg/L to 7.0 mg/L final dissolved oxygen (DO) are excluded from calculations for biochemical oxygen demand (BOD). Disregard bottles 1 and 4 since the final DO for these bottles is outside of the typically accepted range of 2.0 mg/L to 7.0 mg/L.

$DO_{t=0}$	dissolved oxygen concentration at $t = 0$ d	mg/L
$DO_{t=5}$	dissolved oxygen concentration at $t = 5$ d	mg/L
t	time	d
V	volume	mL

For each bottle,

$$\frac{DO_{t=0} - DO_{t=5}}{\dfrac{\text{sample } V}{\text{total } V}}$$

Assume a typical BOD bottle total volume is 300 mL.

For bottle 2,

$$\frac{9.1 \, \frac{mg}{L} - 2.9 \, \frac{mg}{L}}{\dfrac{10 \text{ mL}}{300 \text{ mL}}} = 186 \text{ mg/L}$$

For bottle 3,

$$\frac{9.1 \, \frac{mg}{L} - 6.1 \, \frac{mg}{L}}{\dfrac{5 \text{ mL}}{300 \text{ mL}}} = 180 \text{ mg/L}$$

The BOD_5 at 20°C is

$$\frac{186 \, \frac{mg}{L} + 180 \, \frac{mg}{L}}{2} = 183 \text{ mg/L}$$

k	rate constant	d^{-1}
BOD_u	ultimate BOD	mg/L
BOD_t	BOD at time t	mg/L

$$BOD_u = \frac{BOD_t}{1 - 10^{-kt}}$$
$$= \frac{183 \, \frac{mg}{L}}{1 - 10^{(-0.17/d)(5 \text{ d})}}$$
$$= 213 \text{ mg/L}$$

k_T	rate constant at temperature of interest	d^{-1}
k_{20}	rate constant at 20°C	d^{-1}
T	temperature	°C
θ	temperature correction coefficient	–

Assume a typical value of 1.047 for the temperature correction coefficient.

$$k_T = k_{20} \theta^{T-20}$$
$$k_{15} = \left(\frac{0.17}{d}\right)(1.047^{15-20}) = 0.135 \text{ d}^{-1}$$

The BOD_5 at 15°C is

$$\left(213 \, \frac{mg}{L}\right)\left(1 - 10^{(-0.135/d)(5 \text{ d})}\right)$$
$$= 168 \text{ mg/L} \quad (170 \text{ mg/L})$$

The answer is (C).

Why Other Options Are Wrong

(A) This incorrect solution improperly applies the ultimate BOD equation. Other assumptions, definitions, and equations are unchanged from the correct solution.

(B) This incorrect solution did not disregard bottles 1 and 4. Other assumptions, definitions, and equations were unchanged from the correct solution.

(D) This incorrect solution does not correct for temperature. Other assumptions, definitions, and equations are unchanged from the correct solution.

SOLUTION 79

When the initial pH is less than 8.3, the bicarbonate alkalinity and total alkalinity are equal. 1 mL of 0.03 N H_2SO_4 will titrate 1.5 mg alkalinity as $CaCO_3$.

M	bicarbonate alkalinity	mg/L as $CaCO_3$
M_{total}	total alkalinity	mg/L as $CaCO_3$
V_{sample}	sample volume	mL
$V_{titrant}$	titrant volume	mL
W	acid equivalent as $CaCO_3$	mg as $CaCO_3$/ mL acid

$$M = M_{total} = \frac{WV_{titrant}}{V_{sample}}$$

$$= \frac{(1.5 \text{ mg alk as CaCO}_3)(14.5 \text{ mL } 0.03 \text{ N H}_2\text{SO}_4) \times \left(1000 \dfrac{\text{mL}}{\text{L}}\right)}{(1 \text{ mL } 0.03 \text{ N H}_2\text{SO}_4)(500 \text{ mL sample})}$$

$$= 44 \text{ mg/L as CaCO}_3$$

The answer is (C).

Why Other Options Are Wrong

(A) This incorrect choice calculates alkalinity correctly, but then attempts to correct it for a 500 mL sample. The concentration is independent of the sample size. Other assumptions, definitions, and equations are unchanged from the correct solution.

(B) This incorrect choice assumes that 1 mL of the 0.03 N H_2SO_4 titrant would neutralize 1 mg of alkalinity as $CaCO_3$. This is only true if the standard titrant concentration of 0.02 N H_2SO_4 is used. Other assumptions, definitions, and equations are unchanged from the correct solution.

(D) This incorrect solution calculates alkalinity in mg/L as HCO_3^- and then converts units to mg/L as $CaCO_3$. The titrant used gives alkalinity units in mg as $CaCO_3$, and no subsequent conversion is required. Other assumptions and definitions are unchanged from the correct solution.

SOLUTION 80

Assume that the water temperature along the river course is constant and that the water salinity is low (fresh water).

The saturated dissolved oxygen concentration at 8.6°C in fresh water is 11.7 mg/L.

D_o	dissolved oxygen (DO) deficit at discharge point	mg/L
DO_i	DO concentration at discharge point	mg/L
DO_s	saturated DO concentration	mg/L

$$D_o = DO_s - DO_i = 11.7 \dfrac{\text{mg}}{\text{L}} - 9.3 \dfrac{\text{mg}}{\text{L}}$$
$$= 2.4 \text{ mg/L}$$

k_d	deoxygenation rate constant	d^{-1}
k_r	reoxygenation rate constant	d^{-1}
L_o	ultimate biochemical oxygen demand (BOD$_u$) at the discharge point	mg/L
t_c	critical time	d

$$t_c = \left(\dfrac{1}{k_r - k_d}\right) \ln\left(\left(\dfrac{k_r}{k_d}\right)\left(1 - \dfrac{D_o(k_r - k_d)}{k_d L_o}\right)\right)$$

$$= \left(\dfrac{0.5}{d} - \dfrac{0.4}{d}\right)^{-1}$$

$$\times \ln\left(\left(\dfrac{\dfrac{0.5}{d}}{\dfrac{0.4}{d}}\right)\left(1 - \dfrac{\left(2.4 \dfrac{\text{mg}}{\text{L}}\right)\left(\dfrac{0.5}{d} - \dfrac{0.4}{d}\right)}{\left(\dfrac{0.4}{d}\right)\left(9.8 \dfrac{\text{mg}}{\text{L}}\right)}\right)\right)$$

$$= 1.6 \text{ d}$$

d_c distance to the monitoring point at t_c mi

$$d_c = (1.6 \text{ days})\left(0.3 \dfrac{\text{ft}}{\text{sec}}\right)\left(86{,}400 \dfrac{\text{sec}}{\text{day}}\right)\left(\dfrac{1 \text{ mi}}{5280 \text{ ft}}\right)$$

$$= 7.85 \text{ mi} \quad (7.9 \text{ mi})$$

The answer is (C).

Why Other Options Are Wrong

(A) This incorrect choice fails to invert the rate constant difference in the first term. Other assumptions, definitions, and equations are unchanged from the correct solution.

(B) This incorrect choice uses the actual dissolved oxygen concentration at the discharge point instead of the oxygen deficit to calculate critical time. Other assumptions, definitions, and equations are unchanged from the correct solution.

(D) This incorrect choice confuses dissolved oxygen deficit and ultimate biochemical oxygen demand when calculating the critical time. Other assumptions, definitions, and equations are unchanged from the correct solution.

SOLUTION 81

Environmental Protection Agency exposure factors for ingestion of soil by children are

BW	body weight	15 kg
DI	daily intake	200 mg

Calculate the average daily dose.

ADD	average daily dose	mg/kg·d
C	concentration	ppb or μg/kg

$$\text{ADD} = \frac{C(\text{DI})}{\text{BW}}$$

$$\text{ADD}_{\text{As}} = \frac{\left(1.3\ \frac{\mu g}{\text{kg}}\right)\left(\frac{1\ \text{kg}}{10^9\ \mu g}\right)\left(200\ \frac{\text{mg}}{\text{d}}\right)}{15\ \text{kg}}$$

$$= 0.000\,000\,017\,3\ \text{mg/kg·d}$$

HQ	hazard quotient	–
RfD	reference dose	mg/kg·d

$$\text{HQ} = \frac{\text{ADD}}{\text{RfD}}$$

$$\text{HQ}_{\text{As}} = \frac{0.000\,000\,017\,3\ \frac{\text{mg}}{\text{kg·d}}}{0.0003\ \frac{\text{mg}}{\text{kg·d}}} = 0.000\,058$$

$$\text{ADD}_{\text{Cd}} = \frac{\left(0.96\ \frac{\mu g}{\text{kg}}\right)\left(\frac{1\ \text{kg}}{10^9\ \mu g}\right)\left(200\ \frac{\text{mg}}{\text{d}}\right)}{15\ \text{kg}}$$

$$= 0.000\,000\,012\,8\ \text{mg/kg·d}$$

$$\text{HQ}_{\text{Cd}} = \frac{0.000\,000\,012\,8\ \frac{\text{mg}}{\text{kg·d}}}{0.0005\ \frac{\text{mg}}{\text{kg·d}}} = 0.000\,026$$

$$\text{ADD}_{\text{F}} = \frac{\left(0.42\ \frac{\mu g}{\text{kg}}\right)\left(\frac{1\ \text{kg}}{10^9\ \mu g}\right)\left(200\ \frac{\text{mg}}{\text{d}}\right)}{15\ \text{kg}}$$

$$= 0.000\,000\,005\,6\ \text{mg/kg·d}$$

$$\text{HQ}_{\text{F}} = \frac{0.000\,000\,005\,6\ \frac{\text{mg}}{\text{kg·d}}}{0.0003\ \frac{\text{mg}}{\text{kg·d}}} = 0.000\,019$$

HI   hazard index   –

$$\text{HI} = \sum \text{HQ} = 0.000\,058 + 0.000\,026 + 0.000\,019$$
$$= 0.000\,10$$

The answer is (C).

Why Other Options Are Wrong

(A) This incorrect option calculated chronic daily intake, which applies to carcinogens, instead of average daily dose, which applies to noncarcinogens, and uses the reference dose as a potency factor. The result is risk, not hazard index. This required an assumption that the units given for reference dose should be $(\text{mg/kg·d})^{-1}$ instead of mg/kg·d. Other assumptions, definitions, and equations are unchanged from the correct solution.

(B) This incorrect option calculated chronic daily intake instead of average daily dose. Other assumptions, definitions, and equations are unchanged from the correct solution.

(D) This incorrect option uses parts per billion as a ratio of mass to volume instead of mass to mass and uses 1 L for the daily intake. Other assumptions, definitions, and equations are unchanged from the correct solution.

SOLUTION 82

Algal growth is stimulated by excess nutrients, such as nitrogen, that may be discharged with wastewater effluents. Nitrification and denitrification are commonly applied to control excess nutrients in wastewater discharges and reduce their influence on algal growth in receiving waters. Microscreening may be used to physically remove algae as a final step in wastewater treatment, and is especially suitable where treatment involves stabilization ponds. It may also be appropriate in some cases to apply chlorination to deactivate algae and other organisms and to remove excess nutrients by oxidizing ammonia. Aeration would likely have little effect as an algae control strategy, but may be effective for materials other than algae that are often associated with taste and odor.

The answer is (A).

Why Other Options Are Wrong

(B) This is an incorrect choice because microscreening is applied as a physical process for removing algae from stabilization pond and other wastewater treatment process effluents.

(C) This is an incorrect choice because nitrification and denitrification are routinely employed to remove nitrogen from wastewater effluents. Excess nitrogen in wastewater discharges is commonly associated with algal blooms in receiving waters.

(D) This is an incorrect choice because chlorination can be successfully applied to directly deactivate algae as well as to indirectly control algae by oxidizing ammonia. For wastewater effluents, chlorination will usually be followed by dechlorination.

SOLUTION 83

DO_i dissolved oxygen concentration
          at the discharge   mg/L

$$\text{DO}_i = 10.9\ \frac{\text{mg}}{\text{L}} - 3.2\ \frac{\text{mg}}{\text{L}} = 7.7\ \text{mg/L}$$

By observation, options (B) and (D) do not represent the stream's dissolved oxygen profile below the discharge. Option (B) shows that the dissolved oxygen concentration at the discharge ($t = 0$ d) is not near 7.7 mg/L, as it should be. Also, option (B) is not consistent with the typical profile for an oxygen sag curve. Option (D) shows the dissolved oxygen concentration after about 11 days to be greater than the saturated dissolved oxygen concentration of 10.9 mg/L. The saturated dissolved oxygen concentration is the maximum value possible in the stream under the conditions given.

D_o	dissolved oxygen deficit at the discharge	mg/L
k_d	deoxygenation rate constant	d^{-1}
k_r	reaeration rate constant	d^{-1}
L_o	BOD_u at discharge point	mg/L
t_c	time of critical oxygen sag point	d

$$t_c = (k_r - k_d)^{-1} \ln\left(\frac{k_r}{k_d}\right)\left(1 - \frac{D_o(k_r - k_d)}{k_d L_o}\right)$$

$$= \left(\frac{0.07}{d} - \frac{0.04}{d}\right)^{-1}$$

$$\times \ln\left(\frac{\frac{0.07}{d}}{\frac{0.04}{d}}\left(1 - \frac{\left(3.2\,\frac{mg}{L}\right)\left(\frac{0.07}{d} - \frac{0.04}{d}\right)}{\left(\frac{0.04}{d}\right)\left(7.2\,\frac{mg}{L}\right)}\right)\right)$$

$$= 5.1/d$$

Comparing options (A) and (C) to the critical oxygen sag point, only in option (A) does the curve match the critical time of 5.1 d. Option (C) is not representative of the stream below the discharge.

The answer is (A).

Why Other Options Are Wrong

(B) This illustration is incorrect because the dissolved oxygen deficit for the illustration does not match the calculated value. From the correct solution, DO_i is 7.7 mg/L. Option (B) shows that the dissolved oxygen concentration at the discharge ($t = 0$ d) is near 3.2 mg/L, not the correct concentration of 7.7 mg/L. Option (B) is not consistent with the typical profile for an oxygen sag curve.

(C) This illustration is incorrect because the critical time on the curve does not match the calculated value. From the correct solution, t_c is 5.1 d. Comparing (C) to the critical oxygen sag point, the curve does not match the critical time of 5.1 d and is not representative of the stream below the discharge.

(D) This illustration is incorrect because it shows the dissolved oxygen concentration after about 11 days to be greater than the saturated dissolved oxygen concentration of 10.9 mg/L. The saturated dissolved oxygen concentration is the maximum value possible in the stream under the conditions given, so the illustration does not represent the stream below the discharge.

SOLUTION 04

Using the illustration, the river equilibrium dissolved oxygen concentration after full recovery is 10.5 mg/L. At 65% of recovery, the concentration is

$$\left(10.5\,\frac{mg}{L}\right)(0.65) = 6.8\,mg/L$$

Again using the illustration, a dissolved oxygen concentration of 6.8 mg/L and three days correspond to a mixed-flow BOD_u concentration of 14 mg/L.

Effluent flow is

$$\left(12 \times 10^6\,\frac{gal}{day}\right)\left(0.134\,\frac{ft^3}{gal}\right)\left(\frac{1\,day}{86{,}400\,sec}\right)$$
$$= 18.6\,ft^3/sec$$

S wastewater BOD_u mg/L

For complete mixing downstream of the discharge,

$$\left(37\,\frac{ft^3}{sec}\right)\left(6\,\frac{mg}{L}\right) + \left(18.6\,\frac{ft^3}{sec}\right)S$$
$$= \left(37\,\frac{ft^3}{sec} + 18.6\,\frac{ft^3}{sec}\right)\left(14\,\frac{mg}{L}\right)$$

$$S = \frac{\left(37\,\frac{ft^3}{sec} + 18.6\,\frac{ft^3}{sec}\right)\left(14\,\frac{mg}{L}\right) - \left(37\,\frac{ft^3}{sec}\right)\left(6\,\frac{mg}{L}\right)}{18.6\,\frac{ft^3}{sec}}$$

$$= 30\,mg/L$$

The answer is (B).

Why Other Options Are Wrong

(A) This incorrect solution uses the mixed-flow BOD_u, taken from the illustration, for the allowable wastewater BOD.

(C) This incorrect solution misreads 12 MGD as 12 ft^3/sec. Other definitions are unchanged from the correct solution.

(D) This incorrect solution uses the recovered dissolved oxygen concentration as the dissolved oxygen concentration at the discharge. Other definitions are unchanged from the correct solution.

SOLUTION 85

Gasoline that has leaked from an underground tank and is later discovered in a monitoring well will experience weathering. Weathering represents a loss of lighter fractions to evaporation and other physical and chemical changes that creates on the chromatograph an increase in the proportion of heavier fractions.

The lighter fractions occur earlier and the heavier fractions occur later on the chromatograph. For a given gasoline, the timing of the peaks will be the same between fresh and weathered gasoline, but their relative magnitude will change.

Comparing the choices with the illustration given in the problem statement, option (A) is identical and does not represent any changes due to weathering or other phenomena. Option (B) shows peaks occurring at different times than the problem statement illustration, indicating a different gasoline. Option (D) is enriched in lighter fractions, showing the opposite affect of weathering. Option (C) shows the peaks occurring at the same time as the problem statement illustration with enrichment in the heavier fractions.

The answer is (C).

Why Other Options Are Wrong

(A) The illustration is identical to the problem statement illustration and does not show any changes due to weathering or other phenomena. Weathering would have occurred between the time of the release from the underground tank and its discovery in the monitoring well.

(B) The illustration shows peaks occurring at different times than those of the problem statement illustration. This indicates that the gasolines represented by the two illustrations are not the same and do not have the underground tank as their common origin.

(D) The illustration shows enrichment in lighter fractions, an unexpected phenomenon that would be inconsistent with a gasoline release. Instead of a relative increase in lighter fractions, a relative increase in heavier fractions would be expected.

WATER TREATMENT

SOLUTION 86

Q	total flow rate	m³/d
Q_B	bypass flow rate	m³/d
Q_T	treatment flow rate	m³/d
TH	initial total hardness	mg/L as CaCO₃
TH_D	desired total hardness	mg/L as CaCO₃
TH_T	treated water total hardness	mg/L as CaCO₃

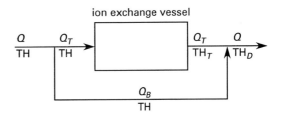

$$Q_T \text{TH}_T + Q_B \text{TH} = Q \text{TH}_D$$

For ion exchange, assume total hardness removal is 100% so that the treated water total hardness is zero.

$$Q_T(0) + Q_B \text{TH} = Q \text{TH}_D$$

$$Q_B = \frac{\text{TH}_D Q}{\text{TH}}$$

$$= \frac{\left(100 \ \frac{\text{mg}}{\text{L}} \text{ as CaCO}_3\right)\left(130\,000 \ \frac{\text{m}^3}{\text{d}}\right)}{382 \ \frac{\text{mg}}{\text{L}} \text{ as CaCO}_3}$$

$$= 34\,031 \ \text{m}^3/\text{d}$$

$$Q_T = Q - Q_B$$

$$= 130\,000 \ \frac{\text{m}^3}{\text{d}} - 34\,031 \ \frac{\text{m}^3}{\text{d}}$$

$$= 95\,969 \ \text{m}^3/\text{d}$$

TH_R total hardness removed   kg/d

$$\text{TH}_R = \text{TH} Q_T$$

$$= \left(382 \ \frac{\text{mg}}{\text{L}} \text{ as CaCO}_3\right)\left(95\,969 \ \frac{\text{m}^3}{\text{d}}\right)$$

$$\times \left((10)^{-6} \ \frac{\text{kg}}{\text{mg}}\right)\left(1000 \ \frac{\text{L}}{\text{m}^3}\right)$$

$$= 36\,660 \ \text{kg/d}$$

C_R exchange resin capacity   kg/m³·resin
V_R exchange resin volume   m³/d

$$V_R = \frac{\text{TH}_R}{C_R} = \frac{36\,660 \ \frac{\text{kg}}{\text{d}}}{95 \ \frac{\text{kg}}{\text{m}^3 \cdot \text{resin}}}$$

$$= 386 \ \text{m}^3 \cdot \text{resin/d}$$

n number of exchange vessels   –
V_C exchange resin volume per vessel   m³

$$n = \frac{V_R}{V_C} = \frac{\left(600 \ \frac{m^3 \cdot resin}{d}\right)\left(1 \ \frac{d}{regeneration}\right)}{4 \ \frac{m^3 \cdot resin}{vessel \cdot regeneration}}$$

$$= 96.5 \text{ vessels} \quad (97 \text{ vessels})$$

The answer is (C).

Why Other Options Are Wrong

(A) This incorrect solution bases the resin use on the bypass flow instead of the treatment flow. Other assumptions, definitions, and equations are the same as used in the correct solution.

(B) This incorrect solution improperly calculated the total hardness removed. Other assumptions, definitions, and equations are the same as used in the correct solution.

(D) This incorrect solution bases the resin use on the total flow instead of correcting for bypass flow to maintain the desired total hardness concentration. Other assumptions, definitions, and equations are the same as used in the correct solution.

SOLUTION 87

Q	flow rate	ft^3/sec
t	residence time	sec
V	mixing basin volume	ft^3

$$V = Qt$$

$$= \left(5.0 \times 10^6 \ \frac{gal}{day}\right)\left(0.134 \ \frac{ft^3}{gal}\right)$$

$$\times (2 \ min)\left(\frac{1 \ day}{1440 \ min}\right)$$

$$= 931 \ ft^3$$

G	velocity gradient	sec^{-1}
μ	dynamic viscosity	$lbf\text{-}sec/ft^2$
P	mixing power	$ft\text{-}lbf/sec$

For water at 60°F, the dynamic viscosity is 2.359×10^{-5} lbf sec/ft².

$$P = G^2 V \mu$$

$$= (700 \ sec^{-1})^2 (931 \ ft^3)\left(2.359 \times 10^{-5} \ \frac{lbf\cdot sec}{ft^2}\right)$$

$$= 10,762 \ ft\text{-}lbf/sec$$

E	motor efficiency	fraction
P_a	mixing power corrected for motor efficiency	hp

$$P_a = \frac{P}{E} = \frac{\left(10,762 \ \frac{ft\cdot lbf}{sec}\right)(1 \ hp)}{\left(550 \ \frac{ft\cdot lbf}{sec}\right)(0.88)} = 22.2 \ hp$$

The standard motor size just greater than 22.2 hp is 25 hp.

The answer is (D).

Why Other Options Are Wrong

(A) This incorrect solution multiplies instead of divides the mixing power by the motor efficiency and selects the standard motor size closest to the required power instead of sizing up. Other assumptions, definitions and equations are unchanged from the correct solution.

(B) This incorrect solution fails to include the correction for motor efficiency. Other assumptions, definitions and equations are unchanged from the correct solution.

(C) This incorrect solution ignores the standard motor size, rounding the required horsepower to the nearest whole number. Other assumptions, definitions and equations are unchanged from the correct solution.

SOLUTION 88

d	basin depth	m
L	basin length per section	m
V	mixing basin volume per section	m^3
w	basin width	m

$$V = Lwd = (4.0 \ m)(4.0 \ m)(8 \ m) = 128 \ m^3$$

D	paddle wheel diameter	m
l_c	paddle-side wall clearance	m

Assume a typically acceptable paddle-wall clearance of 0.3 m.

$$D = d - 2l_c = 4.0 \ m - (2)(0.3 \ m) = 3.4 \ m$$

f_s	paddle slip factor	–
v_p	paddle tip velocity	m/s
ω	paddle rotational speed	rev/min

Assume a typical paddle slip factor of 0.75. Note that πD has units of m/rev.

$$v_p = \omega \pi D f_s$$

$$= \left(3 \ \frac{rev}{min}\right)\pi \left(3.4 \ \frac{m}{rev}\right)(0.75)\left(\frac{1 \ min}{60 \ s}\right)$$

$$= 0.40 \ m/s$$

A_p	paddle surface area facing rotation	m^2

C_D drag coefficient —
G velocity gradient $\sec^{-1}$
μ dynamic viscosity $0.001\,002$ kg/m·s
ρ_w water density 1000 kg/m³

Assume that the drag coefficient for flat paddles is 1.8.

$$A_p = \frac{2G^2 V \mu}{C_D \rho_w \mathrm{v}_p^3}$$

$$= \frac{\left(45\,\frac{1}{\mathrm{s}}\right)^2 (2)(128\text{ m}^3)\left(0.001\,002\,\frac{\mathrm{kg}}{\mathrm{m\cdot s}}\right)}{(1.8)\left(1000\,\frac{\mathrm{kg}}{\mathrm{m}^3}\right)\left(0.40\,\frac{\mathrm{m}}{\mathrm{s}}\right)^3}$$

$$= 4.5 \text{ m}^2$$

The answer is (D).

Why Other Options Are Wrong

(A) This incorrect solution fails to cube the velocity term in the paddle area equation. Other assumptions, definitions, and equations are unchanged from the correct solution.

(B) This incorrect solution fails to include the slip factor when calculating the paddle tip velocity. Other assumptions, definitions, and equations are unchanged from the correct solution.

(C) This incorrect solution does not include the wall clearance when calculating the paddle diameter. Other assumptions, definitions, and equations are unchanged from the correct solution.

SOLUTION 89

The overflow rate is equal to the rise velocity of the water in the sedimentation basin. For those particles, removal will result in a settling velocity that is greater than the overflow rate.

A_s settling zone surface area ft²
q_o overflow rate ft³/ft²-hr
Q flow rate gal/day

$$q_o = \frac{Q}{A_s} = \frac{\left(2.7 \times 10^6 \,\frac{\mathrm{gal}}{\mathrm{day}}\right)\left(0.134\,\frac{\mathrm{ft}^3}{\mathrm{gal}}\right)}{(5700 \text{ ft}^2)\left(24\,\frac{\mathrm{hr}}{\mathrm{day}}\right)}$$

$$= 2.6 \text{ ft}^3/\text{ft}^2\text{-hr}$$

$E\%$ removal efficiency %
v_s particle settling velocity in/sec

For calculating removal efficiency, the overflow rate and particle-settling velocity must have the same units.

$$E\% = \frac{\mathrm{v}_s}{q_o} \times 100\%$$

$$= \frac{\left(0.008\,\frac{\mathrm{in}}{\mathrm{sec}}\right) \times 100\% \left(3600\,\frac{\mathrm{sec}}{\mathrm{hr}}\right)}{\left(2.6\,\frac{\mathrm{ft}^3}{\mathrm{ft}^2\text{-hr}}\right)\left(12\,\frac{\mathrm{in}}{\mathrm{ft}}\right)}$$

$$= 92\%$$

The answer is (D).

Why Other Options Are Wrong

(A) This incorrect solution makes an error when converting from hours to seconds in the efficiency equation. Other assumptions, definitions, and equations are unchanged from the correct solution.

(B) This incorrect solution divides the surface area by the flow rate to calculate efficiency. Other assumptions, definitions, and equations are unchanged from the correct solution.

(C) This incorrect solution uses the wrong conversion from ft³ to gal in the overflow rate calculation. Other assumptions, definitions, and equations are unchanged from the correct solution.

SOLUTION 90

f_P permeate recovery fraction —
Q required feed water flow rate m³/d
Q_o desired fresh water flow rate m³/d

$$Q = \frac{Q_o}{f_p} = \frac{30\,000\,\frac{\mathrm{m}^3}{\mathrm{d}}}{0.77} = 38\,961 \text{ m}^3/\text{d}$$

A_m required membrane area m²
G membrane flux rate m³/m²·d

$$A_m = \frac{Q}{G} = \frac{38\,961\,\frac{\mathrm{m}^3}{\mathrm{d}}}{0.93\,\frac{\mathrm{m}^3}{\mathrm{m}^2\cdot\mathrm{d}}} = 41\,894 \text{ m}^2$$

P_m membrane packing density m²/m³
V_m membrane volume m³

$$V_m = \frac{A_m}{P_m} = \frac{41\,894 \text{ m}^2}{800\,\frac{\mathrm{m}^2}{\mathrm{m}^3}} = 52.4 \text{ m}^3$$

n_M number of membrane modules
V_M membrane module volume m³

$$n_M = \frac{V_m}{V_M} = \frac{52.4 \text{ m}^3}{0.028 \frac{\text{m}^3}{\text{module}}} = 1871 \text{ module}$$

n_R total number of pressure vessels
n_P number of modules per pressure vessel

$$n_R = \frac{n_M}{n_P} = \frac{1871 \text{ module}}{12 \frac{\text{module}}{\text{vessel}}}$$
$$= 156 \text{ vessels} \quad (160 \text{ vessels})$$

The answer is (C).

Why Other Options Are Wrong

(A) This incorrect solution multiplies by instead of divides by the permeate recovery fraction when calculating the required feed water flow rate. Other assumptions, definitions, and equations are unchanged from the correct solution.

(B) This incorrect solution fails to correct the desired fresh water flow rate for permeate recovery. Other assumptions, definitions, and equations are unchanged from the correct solution.

(D) This incorrect solution calculates the required number of modules instead of the required number of pressure vessels. Other assumptions, definitions, and equations are unchanged from the correct solution.

SOLUTION 91

Assume that the Carmen-Kozeny equations apply.

d_g geometric mean particle diameter
  between sieve sizes m
d_{12} mesh opening for a #12 sieve m
d_{16} mesh opening for a #16 sieve m

For a 12 × 16 mesh, assume that 100% of the media passes a #12 sieve and is retained on a #16 sieve. The mesh opening for a #12 sieve is 0.0017 m and for a #16 sieve is 0.001 18 m.

$$d_g = \sqrt{d_{12}d_{16}} = \sqrt{(0.0017 \text{ m})(0.001\,18 \text{ m})}$$
$$= 0.0014 \text{ m}$$

ϕ particle shape factor –
μ water dynamic viscosity kg/m·s
Re Reynolds number –
ρ_w water density, kg/m^3
v_s clean filter filtering velocity m^3/m^2·s

Assume that the particle shape factor for anthracite sand is 0.75, the water density is 998.23 kg/m^3, and the water dynamic viscosity is 0.001 002 kg/m·s.

$$\text{Re} = \frac{\phi \rho_w v_s d_g}{\mu}$$
$$= \frac{(0.75)\left(998.23 \frac{\text{kg}}{\text{m}^3}\right)\left(0.003 \frac{\text{m}^3}{\text{m}^2\cdot\text{s}}\right)(0.0014 \text{ m})}{0.001\,002 \frac{\text{kg}}{\text{m}\cdot\text{s}}}$$
$$= 3.14$$

α media porosity –
f friction factor –

Assume that a typical porosity for granular media is 0.40.

$$f = \frac{(150)(1-\alpha)}{\text{Re}} + 1.75$$
$$= \frac{(150)(1-0.40)}{3.14} + 1.75$$
$$= 30.4$$

g gravitational acceleration 9.81 m/s^2
h head loss m
L filter bed depth m

$$h = \frac{f(1-\alpha)Lv_s^2}{\phi\alpha^3 d_g g}$$
$$= \frac{(30.4)(1-0.40)(0.75 \text{ m})\times\left(0.003 \frac{\text{m}^3}{\text{m}^2\cdot\text{s}}\right)^2\left(100 \frac{\text{cm}}{\text{m}}\right)}{(0.75)(0.40)^3(0.0014 \text{ m})\left(9.81 \frac{\text{m}}{\text{s}^2}\right)}$$
$$= 19 \text{ cm}$$

The answer is (C).

Why Other Options Are Wrong

(A) This incorrect solution makes an error in the friction factor calculation and fails to cube the porosity term in the head loss equation. Other assumptions, definitions, and equations are unchanged from the correct solution.

(B) This incorrect solution squares instead of cubes the porosity term in the head loss equation. Other assumptions, definitions, and equations are unchanged from the correct solution.

(D) This incorrect solution uses the #12 mesh opening as the particle diameter instead of using the geometric average between the two adjacent sieves and makes an error in the friction factor calculation. Other assumptions, definitions, and equations are unchanged from the correct solution.

SOLUTION 92

For trial 1, use an overflow rate of 0.02 m/min and integrate the area to the left of the curve bounded by the overflow rate and the corresponding mass fraction remaining. Each grid element of the integrated area represents 0.0001 m/min.

v	settling velocity	m/min
x	fraction remaining	–
$v_i x_i$	integrated area	m/min

number of grid elements from illustration = 54

$$v_i x_i = \left(0.0001 \ \frac{m}{min}\right)(54)$$
$$= 0.0054 \ m/min$$

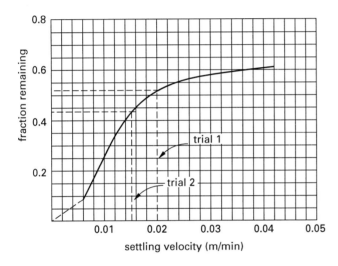

E	fractional efficiency	–
v_c	overflow rate	m/min
x_c	mass fraction remaining corresponding to the overflow rate	–

$$E = (1 - x_c) + \frac{v_i x_i}{v_c}$$
$$= (1 - 0.52) + \frac{0.0054 \ \frac{m}{min}}{0.02 \ \frac{m}{min}}$$
$$= 0.75$$
$$0.75 < 0.80$$

Trial 2 is required.

For trial 2,
$$v_c = 0.015 \ m/min$$

number of grid elements from illustration = 37.5

$$v_i x_i = \left(0.0001 \ \frac{m}{min}\right)(37.5)$$
$$= 0.00375 \ m/min$$
$$E = (1 - x_c) + \frac{v_i x_i}{v_c}$$
$$= (1 - 0.42) + \frac{0.00375 \ \frac{m}{min}}{0.015 \ \frac{m}{min}}$$
$$= 0.83$$

Close enough. Trial 3 is not needed.

The answer is (C).

Why Other Options Are Wrong

(A) This incorrect solution uses the curve to find an overflow rate corresponding to a mass fraction remaining of 0.2. Other assumptions, definitions, and equations are unchanged from the correct solution.

(B) This incorrect solution integrates the area under the curve bounded by the overflow rate. Other assumptions, definitions, and equations are unchanged from the correct solution.

(D) This incorrect solution makes an error in the fractional efficiency equation. The illustration and other assumptions, definitions, and equations are unchanged from the correct solution.

SOLUTION 93

ion	concentration (mg/L)	molecular weight (mg/mmol)	molarity (mmol/L)
Ca^{2+}	187	40	4.7
Mg^{2+}	49	24	2.0
HCO_3^-	618	61	10.1

Assume that the maximum hardness that can be precipitated is 40 mg/L as $CaCO_3$. The molecular weight of $CaCO_3$ is 100 mg/mmol.

CH	HCO_3^- initial	mmol/L
CH_I	HCO_3^- removed	mmol/L
CH_R	HCO_3^- remaining	mmol/L
MgTH	Mg^{2+} hardness initial	mmol/L
$MgTH_I$	Mg^{2+} hardness to be removed	mmol/L
$MgTH_O$	Mg^{2+} hardness precipitated	mmol/L
$MgTH_R$	Mg^{2+} hardness remaining	mmol/L
$NaOH_1$	NaOH added for Ca^{2+} as carbonate hardness removal	mmol/L
$NaOH_2$	NaOH added for Mg^{2+} as carbonate hardness removal	mmol/L

$NaOH_3$ NaOH added for Mg^{2+} as
            non-carbonate removal   mmol/L
TH_D desired total hardness   mmol/L
TH_F final total hardness   mmol/L
TH_R residual total hardness   mmol/L

$$TH_D = \left(100 \frac{\text{mg}}{\text{L}} \text{ as CaCO}_3\right)\left(\frac{1 \text{ mmol}}{100 \text{ mg}} \text{ as CaCO}_3\right)$$
$$= 1.0 \text{ mmol/L}$$

$$TH_R = \left(40 \frac{\text{mg}}{\text{L}} \text{ as CaCO}_3\right)\left(\frac{1 \text{ mmol}}{100 \text{ mg}} \text{ as CaCO}_3\right)$$
$$= 0.4 \text{ mmol/L}$$

$$TH_F = TH_D - TH_R = 1.0 \frac{\text{mmol}}{\text{L}} - 0.4 \frac{\text{mmol}}{\text{L}}$$
$$= 0.6 \text{ mmol/L}$$

$$MgTH_I = MgTH = TH_F = 2.0 \frac{\text{mmol}}{\text{L}} - 0.6 \frac{\text{mmol}}{\text{L}}$$
$$= 1.4 \text{ mmol/L}$$

$$\underset{Ca^{2+}}{(1)\left(4.7 \frac{\text{mmol}}{\text{L}}\right)} + \underset{2HCO_3^-}{(2)\left(4.7 \frac{\text{mmol}}{\text{L}}\right)} + \underset{2NaOH}{(2)\left(4.7 \frac{\text{mmol}}{\text{L}}\right)}$$
$$\rightarrow CaCO_3 \downarrow + 2Na^+ + CO_3^- + H_2O$$

$$CH_R = CH - CH_I = 10.1 \frac{\text{mmol}}{\text{L}} - (2)\left(4.7 \frac{\text{mmol}}{\text{L}}\right)$$
$$= 0.7 \text{ mmol/L}$$

$$NaOH_1 = (2)\left(4.7 \frac{\text{mmol}}{\text{L}}\right) = 9.4 \text{ mmol/L}$$

$$\underset{Mg^{2+}}{\left(\frac{1}{2}\right)\left(0.7 \frac{\text{mmol}}{\text{L}}\right)} + \underset{2HCO_3^-}{\left(\frac{2}{2}\right)\left(0.7 \frac{\text{mmol}}{\text{L}}\right)} + \underset{4NaOH}{\left(\frac{4}{2}\right)\left(0.7 \frac{\text{mmol}}{\text{L}}\right)}$$
$$\rightarrow Mg(OH)_2 \downarrow + 4Na^+ + 2CO_3^- + 2H_2O$$

$$MgTH_R = MgTH_I - MgTH_O$$
$$= 1.4 \frac{\text{mmol}}{\text{L}} - \left(\frac{1}{2}\right)\left(0.7 \frac{\text{mmol}}{\text{L}}\right)$$
$$= 1.05 \text{ mmol/L}$$

$$NaOH_2 = \left(\frac{4}{2}\right)\left(0.7 \frac{\text{mmol}}{\text{L}}\right) = 1.4 \text{ mmol/L}$$

$$\underset{Mg^{2+}}{(1)\left(1.05 \frac{\text{mmol}}{\text{L}}\right)} + \underset{2NaOH}{(2)\left(1.05 \frac{\text{mmol}}{\text{L}}\right)}$$
$$\rightarrow Mg(OH)_2 \downarrow + 2Na^+$$

$$NaOH_3 = (2)\left(1.05 \frac{\text{mmol}}{\text{L}}\right) = 2.1 \text{ mmol/L}$$

f NaOH fractional purity
MW   molecular weight of NaOH   mg/mmol

NaOH   NaOH dose   tonne/mo
Q flow rate   m³/d

$$NaOH = (NaOH_1 + NaOH_2 + NaOH_3) \text{MW} \frac{Q}{f}$$
$$= \left(9.4 \frac{\text{mmol}}{\text{L}} + 1.4 \frac{\text{mmol}}{\text{L}} + 2.1 \frac{\text{mmol}}{\text{L}}\right)$$
$$\times \left(40 \frac{\text{mg}}{\text{mmol}}\right) \times \frac{100\%}{83\%}\left(30\,000 \frac{\text{m}^3}{\text{d}}\right)$$
$$\times \left(30 \frac{\text{d}}{\text{mo}}\right)\left((10)^{-6} \frac{\text{tonne} \cdot \text{L}}{\text{m}^3 \cdot \text{mg}}\right)$$
$$= 560 \text{ tonne/mo}$$

The answer is (B).

Why Other Options Are Wrong

(A) This incorrect solution does not correct for reagent purity. Other assumptions, definitions, and equations are the same as used in the correct solution.

(C) This incorrect solution does not include in the calculation the hardness that is to remain in solution. Other assumptions, definitions, and equations are the same as used in the correct solution.

(D) This incorrect solution assumes that the bicarbonate is unlimited. Other assumptions, definitions, and equations are the same as used in the correct solution.

SOLUTION 94

At an initial pH of 9.7, the dominant alkalinity species will be carbonate.

For hydroxide alkalinity,
$$\text{pH} = 9.7$$
$$\text{pH} + \text{pOH} = 14$$
$$\text{pOH} = 14 - 9.7 = 4.3$$

$[OH^-]$ hydroxide concentration   mol/L

$$\text{pOH} = -\log[OH^-]$$
$$[OH^-] = (10)^{-4.3} \text{ mol/L}$$

EW   calcium carbonate equivalent
            weight   50 000 mg/eq
OH^-_{alk} hydroxide alkalinity   mg/L as CaCO$_3$
$|V|$ valence   eq/mol

$$OH^-_{alk} = [OH^-]|V|EW$$
$$= \left((10)^{-4.3} \frac{\text{mol}}{\text{L}}\right)\left(1 \frac{\text{eq}}{\text{mol}}\right)$$
$$\times \left(50\,000 \frac{\text{mg CaCO}_3}{\text{eq}}\right)$$
$$= 2.5 \text{ mg/L as CaCO}_3$$

When 0.03 N sulfuric acid is used as the titrant, 1.0 mL of acid will neutralize 1.5 mg of alkalinity as $CaCO_3$.

For carbonate alkalinity,

$CO_3^{2-}{}_{alk}$ carbonate alkalinity mg/L as $CaCO_3$
V_{sample} sample volume mL
$V_{titrant}$ titrant volume added mL

Titrating from pH 9.7 to pH 8.3 used 6 mL of titrant,

$$\left(\frac{1}{2}\right)CO_3^{2-}{}_{alk} = \left(\frac{1.5 \text{ mg as } CaCO_3}{1 \text{ mL titrant}}\right)\left(\frac{V_{titrant}}{V_{sample}}\right)$$
$$- OH^-{}_{alk}$$
$$= \frac{(1.5 \text{ mg as } CaCO_3)(6 \text{ mL}) \times \left((10)^3 \frac{\text{mL}}{\text{L}}\right)}{(1 \text{ mL})(250 \text{ mL sample})}$$
$$- 2.5 \frac{\text{mg}}{\text{L}} \text{ as } CaCO_3$$
$$= 33.5 \text{ mg/L as } CaCO_3$$
$$CO_3^{2-}{}_{alk} = (2)\left(33.5 \frac{\text{mg}}{\text{L}} \text{ as } CaCO_3\right)$$
$$= 67 \text{ mg/L as } CaCO_3$$

For total alkalinity,

T_{alk} total alkalinity mg/L as $CaCO_3$

Titrating from pH 9.7 to pH 4.5 used 18 mL of titrant,

$$T_{alk} = \left(\frac{1.5 \text{ mg as } CaCO_3}{1 \text{ mL titrant}}\right)\left(\frac{V_{titrant}}{V_{sample}}\right)$$
$$= \frac{(1.5 \text{ mg as } CaCO_3)(18 \text{ mL})\left((10)^3 \frac{\text{mL}}{\text{L}}\right)}{(1 \text{ mL})(250 \text{ mL sample})}$$
$$= 108 \text{ mg/L as } CaCO_3$$

For bicarbonate alkalinity,

$HCO_3^-{}_{alk}$ bicarbonate alkalinity mg/L as $CaCO_3$

$$HCO_3^-{}_{alk} = T_{alk} - OH^-{}_{alk} - CO_3^{2-}{}_{alk}$$
$$= 108 \frac{\text{mg}}{\text{L}} \text{ as } CaCO_3 - 2.5 \frac{\text{mg}}{\text{L}} \text{ as } CaCO_3$$
$$- 67 \frac{\text{mg}}{\text{L}} \text{ as } CaCO_3$$
$$= 38.5 \text{ mg/L as } CaCO_3$$

Check. Carbonate alkalinity dominates at 67 mg/L as $CaCO_3$.

The answer is (B).

Why Other Options Are Wrong

(A) This incorrect solution does not account for using the non-standard 0.03 N sulfuric acid titrant instead of the standard 0.02 N sulfuric acid titrant. Other assumptions, definitions, and equations are the same as used in the correct solution.

(C) This incorrect solution ignores the hydroxide alkalinity. Other assumptions, definitions, and equations are the same as used in the correct solution.

(D) This incorrect solution assumes that all the carbonate alkalinity is titrated above pH 8.3 instead of one-half of it being titrated. Other assumptions, definitions, and equations are the same as used in the correct solution.

SOLUTION 95

Because decreasing temperature results in a longer reaction time, take 21°C as the reference temperature (T_1) and 17°C as the temperature of interest (T_2).

t time min
T temperature K
E activation energy 6400 cal/mol for aqueous chlorine at pH 8.5
R gas law constant 1.99 cal/mol·K

$$T_1 = (21°C + 273°) = 294K$$
$$T_2 = (17°C + 273°) = 290K$$

$$\ln\frac{t_1}{t_2} = \frac{E(T_2 - T_1)}{RT_1T_2}$$
$$= \frac{\left(6400 \frac{\text{cal}}{\text{mol}}\right)(290K - 294K)}{\left(1.99 \frac{\text{cal}}{\text{mol·K}}\right)(294K)(290K)}$$
$$= -0.15$$
$$t_2 = \frac{t_1}{e^{-0.15}} = \frac{23 \text{ min}}{e^{-0.15}} = 27 \text{ min}$$

The answer is (C).

Why Other Options Are Wrong

(A) This incorrect choice uses °C for temperature instead of K, takes the natural log instead of raising e to the whole-number power, and reverses the temperature values in the equation. Other assumptions, definitions, and equations are the same as for the correct solution.

(B) This incorrect choice reverses the temperature values in the equation. Other assumptions, definitions, and equations are unchanged from the correct solution.

(D) This incorrect choice uses activation energy for chloramines at pH 8.5 instead of for aqueous chlorine. Other assumptions, definitions, and equations are unchanged from the correct solution.

SOLUTION 96

E — permeate recovery fraction —
Q_d — desired flow rate — m^3/d
Q_r — feed flow rate — m^3/d

$$Q_d = \frac{Q_E}{E} = \frac{16\,000\ \frac{m^3}{d}}{0.80} = 20\,000\ m^3/d$$

G_m — membrane flux rate — m^3/m^2·d
V_m — membrane volume — m^3
ρ_m — membrane packing density — m^2/m^3

$$V_m = \frac{Q_d}{G_m \rho_m} = \frac{20\,000\ \frac{m^3}{d}}{\left(0.83\ \frac{m^3}{m^2 \cdot d}\right)\left(800\ \frac{m^2}{m^3}\right)}$$
$$= 30\ m^3$$

The number of pressure vessels is

$$(30\ m^3)\left(\frac{1\ \text{module}}{0.03\ m^3}\right)\left(\frac{1\ \text{pressure vessel}}{10\ \text{modules}}\right)$$
$$= 100\ \text{pressure vessels}$$

The answer is (D).

Why Other Options Are Wrong

(A) This incorrect choice multiplies instead of divides the desired freshwater flow rate by the permeate recovery. Other definitions and equations are unchanged from the correct solution.

(B) This incorrect choice uses the desired freshwater flow rate instead of calculating the feed rate corrected for permeate recovery. Other definitions and equations are unchanged from the correct solution.

(C) This incorrect choice uses the salt rejection percentage instead of the permeate recovery percentage to find the feed rate. Other definitions and equations are unchanged from the correct solution.

ENGINEERING ECONOMICS

SOLUTION 97

The typical average annual daily flow for planning purposes is 165 gal/person-day.

Q — total annual flow — gal/yr

$$Q = (65{,}000\ \text{people})\left(165\ \frac{\text{gal}}{\text{person-day}}\right)\left(365\ \frac{\text{day}}{\text{yr}}\right)$$
$$= 3.9 \times 10^9\ \text{gal/yr}$$

e — annual electricity cost — $

$$e = \left(3.9 \times 10^9\ \frac{\text{gal}}{\text{yr}}\right)\left(\frac{\$0.12}{(10)^3\ \text{gal}}\right)(0.20)$$
$$= \$93{,}600/\text{yr}\quad (\$94{,}000/\text{yr})$$

The answer is (C).

Why Other Options Are Wrong

(A) The incorrect solution assumes an incorrect average annual daily flow of 100 gal/person-day. Definitions are unchanged from the correct solution.

(B) This incorrect solution calculates the annual flow on the basis of a 5 day week instead of a 7 day week. Definitions are unchanged from the correct solution.

(D) This incorrect solution divides instead of multiplies by the percent attributable to electrical power. Definitions are unchanged from the correct solution.

SOLUTION 98

The landfill volume is

$$(0.5)\begin{pmatrix}(1200\ \text{ft} - 80\ \text{ft} - 80\ \text{ft})\\ \times (1600\ \text{ft} - 80\ \text{ft} - 80\ \text{ft})\\ + (1200\ \text{ft})(1600\ \text{ft})\end{pmatrix}(80\ \text{ft})$$
$$= 1.4 \times 10^8\ \text{ft}^3$$

The annual waste mass landfilled is

$$(0.75)(215{,}000\ \text{people})\left(4.6\ \frac{\text{lbm}}{\text{capita day}}\right)\left(365\ \frac{\text{day}}{\text{yr}}\right)$$
$$= 2.7 \times 10^8\ \text{lbm/yr}$$

The annual in-place waste volume landfilled is

$$\frac{2.7 \times 10^8\ \frac{\text{lbm}}{\text{yr}}}{50\ \frac{\text{lbm}}{\text{ft}^3}} = 5.4 \times 10^6\ \text{ft}^3/\text{yr}$$

The annual cover volume is

$$\frac{5.4 \times 10^6 \ \frac{\text{ft}^3}{\text{yr}}}{4.5} = 1.2 \times 10^6 \ \text{ft}^3/\text{yr}$$

The annual landfill total volume is

$$5.4 \times 10^6 \ \frac{\text{ft}^3}{\text{yr}} + 1.2 \times 10^6 \ \frac{\text{ft}^3}{\text{yr}} = 6.6 \times 10^6 \ \text{ft}^3/\text{yr}$$

The landfill operating life is

$$\frac{1.4 \times 10^8 \ \text{ft}^3}{6.6 \times 10^6 \ \frac{\text{ft}^3}{\text{yr}}} = 21 \ \text{yr}$$

The answer is (C).

Why Other Options Are Wrong

(A) This incorrect solution reverses the cover-to-fill ratio. Definitions and equations are unchanged from the correct solution.

(B) This solution is incorrect because both the recycled-to-landfilled ratio and the cover-to-fill ratio are reversed. The landfilled mass is based on the percent recycled instead of the percent landfilled. Definitions and equations are unchanged from the correct solution.

(D) This choice is incorrect because the annual waste mass landfilled is calculated based on percent recycled. Definitions and equations are unchanged from the correct solution.

SOLUTION 99

The transfer station will become more economical than direct-haul when the station's cost—the sum of amortized capital, operating costs, and the corresponding haul costs—is less than the direct-haul cost. This occurs at the breakeven point where the two costs are equal.

C_{dh} direct-haul breakeven cost   \$/ton-min

$$\frac{\$3.87}{\text{ton}} + \left(\frac{\$0.016}{\text{ton-min}}\right)(72 \ \text{min}) = (C_{\text{dh}})(72 \ \text{min})$$

$$C_{\text{dh}} = \frac{\frac{\$3.87}{\text{ton}} + \frac{\$1.15}{\text{ton}}}{72 \ \text{min}} = \$0.070/\text{ton-min}$$

The answer is (C).

Why Other Options Are Wrong

(A) This incorrect choice equates the sum of the transportation costs to the amortized transfer station costs. Other definitions are unchanged from the correct solution.

(B) This incorrect choice uses the cost for direct-haul instead of the cost of the transfer station, and then misapplies the result. Other assumptions and definitions are unchanged from the correct solution.

(D) This incorrect choice uses the cost for direct-haul instead of the transfer station haul cost in the calculation. Other assumptions and definitions are unchanged from the correct solution.

SOLUTION 100

The time required per month to collect the current roll-off boxes, based on a 90 min round trip for each box, is

$$(34 \ \text{locations})\left(3 \ \frac{\text{boxes}}{\text{location}}\right)\left(90 \ \frac{\text{min}}{\text{round trip-box}}\right)$$
$$\times \left(2 \ \frac{\text{round trips}}{\text{mo}}\right)\left(\frac{1 \ \text{hr}}{60 \ \text{min}}\right)$$
$$= 306 \ \text{hr/mo}$$

The dumpsters and compaction trucks will collect the same volume of waste on the same schedule, but the compaction trucks will haul an equivalent uncompacted volume of

$$(12 \ \text{yd}^3 \ \text{compacted})\left(\frac{3 \ \text{yd}^3 \ \text{uncompacted}}{1 \ \text{yd}^3 \ \text{compacted}}\right) = 36 \ \text{yd}^3$$

To provide the same dumpster capacity as roll-off box capacity, each location will have

$$\frac{\left(18 \ \frac{\text{yd}^3}{\text{box}}\right)\left(3 \ \frac{\text{boxes}}{\text{location}}\right)}{6 \ \frac{\text{yd}^3}{\text{dumpster}}} = 9 \ \text{dumpsters/location}$$

The total dumpsters emptied per load will be

$$\frac{36 \ \frac{\text{yd}^3}{\text{load}}}{6 \ \frac{\text{yd}^3}{\text{dumpster}}} = 6 \ \text{dumpsters/load}$$

For each location, one truck will complete one round trip to service six dumpsters.

$$(34 \ \text{locations})\left(2 \ \frac{\text{round trips}}{\text{mo}}\right)$$
$$\times \left(90 \ \frac{\text{min}}{\text{round trip}}\right)\left(\frac{1 \ \text{hr}}{60 \ \text{min}}\right)$$
$$= 102 \ \text{hr/mo}$$

Another truck will complete one round trip with one trip between locations to service three dumpsters at that location and three dumpsters at another location.

$$(34 \text{ locations})\left(\frac{2 \text{ round trips}}{2 \text{ locations}}\right)$$
$$\times \left(\frac{90 \text{ min}}{\text{round trip}} + \frac{27 \text{ min}}{\text{round trip}}\right)\left(\frac{1 \text{ hr}}{60 \text{ min}}\right)$$
$$= 66 \text{ hr}$$

The total time per month using the proposed dumpsters and compaction trucks will be

$$102 \frac{\text{hr}}{\text{mo}} + 66 \frac{\text{hr}}{\text{mo}} = 168 \text{ hr/mo}$$

The total driving time saved is

$$306 \frac{\text{hr}}{\text{mo}} - 168 \frac{\text{hr}}{\text{mo}} = 138 \text{ hr/mo} \quad (140 \text{ hr/mo})$$

The answer is (D).

Why Other Options Are Wrong

(A) This incorrect solution includes only one instead of three roll-off boxes per location.

(B) This incorrect solution considers the time for a one-way instead of a two-way trip.

(C) This incorrect solution includes only one pick-up per month instead of the scheduled two.